MINIMALLY INVASIVE MEDICAL TECHNOLOGY

T0187716

Series in Medical Physics

Series Editors:

CG Orton, Karmanos Cancer Institute and Wayne State University
J A E Spaan, University of Amsterdam, The Netherlands
J G Webster, University of Wisconsin-Madison, USA

Other books in the series

The Physics of Medical Imaging
S Webb (ed)

The Physics of Three-Dimensional Radiation Therapy: Conformal Radiotherapy, Radiosurgery and Treatment Planning
S Webb

The Physics of Conformal Radiotherapy: Advances in Technology
S Webb

Intensity-Modulated Radiation Therapy
S Webb

Medical Physics and Biomedical Engineering
B H Brown, R H Smallwood, D C Barber, P V Lawford and D R Hose

Biomedical Magnetic Resonance Technology
C-N Chen and D I Hoult

Rehabilitation Engineering Applied to Mobility and Manipulation
R A Cooper

Physics for Diagnostic Radiology, second edition
P P Dendy and B H Heaton

Linear Accelerators for Radiation Therapy, second edition
D Greene and P C Williams

Health Effects of Exposure to Low-Level Ionizing Radiation
W R Hendee and F M Edwards (eds)

Monte Carlo Calculations in Nuclear Medicine
M Ljungberg, S-E Strand and M A King (eds)

Introductory Medical Statistics, third edition
R F Mould

The Design of Pulse Oximeters
J G Webster (ed)

Ultrasound in Medicine
F A Duck, A C Baker and H C Starritt (eds)

Other titles of interest

Prevention of Pressure Sores: Engineering and Clinical Aspects
J G Webster

Series in Medical Physics

MINIMALLY INVASIVE MEDICAL TECHNOLOGY

Edited by John G Webster

Department of Biomedical Engineering,
University of Wisconsin,
Madison, Wisconsin, USA

CRC Press
Taylor & Francis Group
Boca Raton   London   New York

CRC Press is an imprint of the
Taylor & Francis Group, an **informa** business

First published 2001 by IOP Publishing Ltd

Published 2019 by CRC Press
Taylor & Francis Group
6000 Broken Sound Parkway NW, Suite 300
Boca Raton, FL 33487-2742

First issued in paperback 2019

No claim to original U.S. Government works

ISBN 13: 978-0-367-45541-5 (pbk)
ISBN 13: 978-0-7503-0733-8 (hbk)

Visit the Taylor & Francis Web site at
http://www.taylorandfrancis.com

and the CRC Press Web site at
http://www.crcpress.com

British Library Cataloguing-in-Publication Data

A catalogue record for this book is available from the British Library.

Library of Congress Cataloging-in-Publication Data are available

Series Editors:
C G Orton, Karmanos Cancer Institute and Wayne State University, Detroit, USA
J A E Spaan, University of Amsterdam, The Netherlands
J G Webster, University of Wisconsin-Madison, USA

Cover Design: Victoria Le Billon

Typeset by the Editor using Microsoft Word 2000

The Series in Medical Physics is the official book series of the International Federation for Medical and Biological Engineering (IFMBE) and the International Organization for Medical Physics (IOMP).

IFMBE

The IFMBE was established in 1959 to provide medical and biological engineering with an international presence. The Federation has a long history of encouraging and promoting international cooperation and collaboration in the use of technology for improving the health and life quality of man.

The IFMBE is an organization that is mostly an affiliation of national societies. Transnational organizations can also obtain membership. At present there are 42 national members, and one transnational member with a total membership in excess of 15 000. An observer category is provided to give personal status to groups or organizations considering formal affiliation.

Objectives

• To reflect the interests and initiatives of the affiliated organizations.

• To generate and disseminate information of interest to the medical and biological engineering community and international organizations.

• To provide an international forum for the exchange of ideas and concepts.

• To encourage and foster research and application of medical and biological engineering knowledge and techniques in support of life quality and cost-effective health care.

• To stimulate international cooperation and collaboration on medical and biological engineering matters.

• To encourage educational programmes which develop scientific and technical expertise in medical and biological engineering.

Activities

The IFMBE has published the journal *Medical and Biological Engineering and Computing* for over 34 years. A new journal *Cellular Engineering* was established in 1996 in order to stimulate this emerging field in biomedical engineering. In *IFMBE News* members are kept informed of the developments in the Federation. *Clinical Engineering Update* is a publication of our division of Clinical Engineering. The Federation also has a division for Technology Assessment in Health Care.

Every three years, the IFMBE holds a World Congress on Medical Physics and Biomedical Engineering, organized in cooperation with the IOMP and the IUPESM. In addition, annual, milestone, regional conferences are organized in different regions of the world, such as the Asia Pacific, Baltic, Mediterranean, African and South American regions.

The administrative council of the IFMBE meets once or twice a year and is the steering body for the IFMBE. The council is subject to the rulings of the General Assembly which meets every three years.

For further information on the activities of the IFMBE, please contact Jos A E Spaan, Professor of Medical Physics, Academic Medical Centre, University of Amsterdam, PO Box 22660, Meibergdreef 9, 1105 AZ, Amsterdam, The Netherlands. Tel: 31 (0) 20 566 5200. Fax: 31 (0) 20 691 7233. Email: IFMBE@amc.uva.nl. WWW: http://vub.vub.ac.be/~ifmbe.

IOMP

The IOMP was founded in 1963. The membership includes 64 national societies, two international organizations and 12 000 individuals. Membership of IOMP consists of individual members of the Adhering National Organizations. Two other forms of membership are available, namely Affiliated Regional Organization and Corporate Members. The IOMP is administered by a Council, which consists of delegates from each of the Adhering National Organization; regular meetings of Council are held every three years at the International Conference on Medical Physics (ICMP). The Officers of the Council are the President, the Vice-President and the Secretary-General. IOMP committees include: developing countries, education and training; nominating; and publications.

Objectives

• To organize international cooperation in medical physics in all its aspects, especially in developing countries.
• To encourage and advise on the formation of national organizations of medical physics in those countries which lack such organizations.

Activities

Official publications of the IOMP are *Physiological Measurement*, *Physics in Medicine and Biology* and the *Series in Medical Physics*, all published by Institute of Physics Publishing. The IOMP publishes a bulletin *Medical Physics World* twice a year.
Two Council meetings and one General Assembly are held every three years at the ICMP. The most recent ICMPs were held in Kyoto, Japan (1991), Rio de Janeiro, Brazil (1994), Nice, France (1997) and Chicago, USA (2000). These conferences are normally held in collaboration with the IFMBE to form the World Congress on Medical Physics and Biomedical Engineering. The IOMP also sponsors occasional international conferences, workshops and courses.

For further information contact: Hans Svensson, PhD, DSc, Professor, Radiation Physics Department, University Hospital, 90185 Umeå, Sweden. Tel: (46) 90 785 3891. Fax: (46) 90 785 1588. Email: Hans.Svensson@radfys.umu.se.

CONTENTS

PREFACE

Minimally invasive medicine has the goal of providing health care with minimal trauma. When minimally invasive surgery is utilized, it reduces length of stay, lowers costs, lowers pain and reduces blood loss. Other minimally invasive techniques minimize radiation exposure, tissue damage and drug side effects.

This book emphasizes the technology required to accomplish minimally invasive medicine. It should be of interest to biomedical engineers, medical physicists and health care providers who want to know the technical workings of their devices and instruments.

Chapter 1 describes sensors that are placed on the surface of the body or in miniature catheters to sense parameters that formerly required the removal of blood. Such parameters are hemoglobin oxygenation, partial pressure of carbon dioxide, glucose concentration, and others. Chapter 2 explains integrated circuit microelectrode arrays that minimize the number of electrodes placed into the brain, and also sieve electrodes that promote the connection between nerves and electronic circuits. Chapter 3 details the operation of catheter-tip pressure sensors using fiber-optic or strain-gage principles.

Although X rays are inherently less invasive than surgery, chapter 4 shows how X-ray dose can be minimized using image intensifiers or computed tomography (CT). Chapter 5 shows how nuclear medicine techniques of single-photon emission computed tomography and positron emission tomography provide functional information with minimal invasion. Chapter 6 describes how magnetic resonance imaging (MRI) provides excellent images of anatomy and function with no ionizing radiation.

Chapter 7 explains how bioelectric and biomagnetic functional data can be combined with anatomic images. Chapter 8 details the transducers and signal processing for creating noninvasive ultrasound images. Multimodal imaging in chapter 9 is an imaging method where multiple distinct anatomical imaging modalities as well as functional information are fused into a single framework.

Chapter 10 describes minimally invasive surgery, in which the site is accessed via small incisions and advanced surgical tools are employed to perform cutting, coagulation and vaporization with minimal injury to the surrounding tissue. Chapter 11 shows how rigid and flexible endoscopes provide images for laparoscopic and arthroscopic surgery.

Chapter 12 explains how the surgeon can benefit from image-guided surgery using images combined from CT, MRI, X ray, ultrasound, etc. Chapter 13 shows how virtual reality environments can create three-dimensional visual and sound perception and tactile feedback to guide interactive surgery. Robotics can augment surgery in planning and feedback, precision, access and dexterity, accuracy, and specialization and access to care. Chapter 14 describes robots used in neurosurgery and eye, ear, gastrointestinal and joint surgery.

Chapter 15 details ablation—destruction of tissue—by heat, cold, laser, ultrasound, or chemicals to avoid more destructive knife surgery. Chapter 16 describes electric stimulation of the lung, brain, bladder and heart (cardiac pacemakers) to replace lost function. Chapter 17 describes the modulation of beam intensity by multi-leaf collimators during radiotherapy to optimally spare healthy tissue while destroying tumors. Chapter 18 covers inhalational, transdermal, and gastrointestinal drug administration that minimizes systemic effects and increases localized effects.

We provide problems as a means of provoking further thought toward learning the given information. Rather than giving an exhaustive list of references, we have included review articles and books that can serve as an entry into further study. All contributors are from the University of Wisconsin, Madison, WI, USA, and worked as a team to write this book. We would welcome suggestions for improvement of subsequent printings and editions.

John G Webster
Department of Biomedical Engineering
University of Wisconsin–Madison
Madison WI 53706
USA
webster@engr.wisc.edu
December 2000

CHAPTER 1

CHEMICAL SENSORS

Hong Cao

1.1 OBJECTS OF MEASUREMENT

1.1.1 Objects of chemical measurement

Chemical analysis measures the chemical substance concentration inside the human body to provide information for diagnosis and therapy. Usually the chemical concentration inside the body is estimated by analyzing a sampled material such as blood, urine, or cerebrospinal fluid, as well as other fluids. However, continuous monitoring *in vivo* is required in some circumstances, such as gas monitoring during anesthesia, long-term glucose monitoring for diabetes and rapidly changing chemical concentrations.

Blood is the most common object for chemical measurement, because the blood circulation transports many important chemical substances. The concentrations of many substances in the blood reflect those in the whole body. Conventionally, blood samples are taken from a vein and analyzed by a chemical analyzer in a clinical laboratory. During continuous monitoring, either drainage of the blood or placing a chemical sensor in a blood vessel is required.

Urine is also a common object in clinical laboratory analysis. Although the information available from urine is less than that from blood, urine analysis is advantageous because the sample is easier to obtain, especially for drainage during continuous monitoring.

The common analyzed chemical substances are glucose, protein, hemoglobin, acidity (H^+), oxygen (O_2) and carbon dioxide (CO_2). The amount of substance is expressed by mass concentration (mg ml^{-1}, g ml^{-1}) or molar concentration (mol l^{-1}, mmol l^{-1}, etc). For gas dissolved in blood, it is more common to use the partial pressure to express the concentration, which is denoted as pO_2 and pCO_2 (Pa or mmHg). And for concentration of a hydrogen ion, the most used unit is the pH value, which is

$$pH = -\log_{10}[H^+] \qquad (1.1)$$

where H$^+$ is expressed in mol l^{-1}.

1.1.2 Requirement of chemical-measurement sensor

Chemical sensors are used in various clinical situations, such as anesthesia, intensive care, emergency operations, and ambulatory monitoring of postoperative and chronic patients. It is also an important component of a closed-loop artificial organ, in which chemical quantities inside the human body are servo-controlled to a desired level.

Chemical sensors can be applied to the human body using different approaches. These methods include attaching the sensor to the surface of the human body, inserting it into the tissue or blood vessel, implanting it inside the body, attaching it to extracorporeal circulation, or installing it in an artificial organ. When these chemical sensors are used *in vivo*, we must consider the following requirements for performance and safety:

1.   long-term stability if the sensor is implanted inside the human body;
2.   sterilization during initial calibration and recalibration;
3.   the chemical reaction occurring at the sensor should not cause toxicity to the tissue;
4.   when the sensor is placed in a blood vessel, it should not cause coagulation or obstruct the blood flow;
5.   the sensor should be electrically insulated from the body fluid so that the leakage current never causes an electric hazard;
6.   the geometry and material of the sensor should not cause any mechanical hazard to the surrounding tissue.

1.1.3 Placement of sensors

When monitoring chemical quantities inside the body, a sensor can be inserted into the body at the location of measurement, which is called *in vivo* measurement. Or the body fluid can be drained or sucked continuously from the body and the measurement performed by a sensor outside the body, which is called *ex vivo* measurement. These procedures are all invasive. To reduce the invasiveness (or to be minimally invasive), we want the sensor size as small as possible to reduce the pain when it is inserted.

If the measured substance is permeable through skin, the measurement can be performed through the skin, which is noninvasive. Unfortunately, the skin is impermeable to most clinically important substances. Only some gases such as oxygen and carbon dioxide are fairly permeable and their partial pressure can be measured transcutaneously. When the outermost layer of the skin, the stratum corneum, is removed, interstitial fluid can be sucked out and some substances such as glucose can then be measured.

1.2 ELECTROCHEMICAL SENSORS

1.2.1 Electrode potential

Electrochemical sensors convert a chemical quantity into electric potential or electric current, based on electrochemical principles. Figure 1.1 shows the fundamental electrochemical measurement system. It consists of two electrodes (one as a reference electrode), electrolyte solution and the electronic circuits. When the electrolyte is dissolved in solution, it dissociates into ions. If an electric field is applied in the solution, the ions move in the electrolyte and form an electric current in the solution. However, the electronic circuit cannot measure the ion current directly, because the current in the circuit is due to the movement of the free electrons instead of chemical ions. Hence a pair of electrodes converts the ionic current into the electric current. In electrochemistry, an electrode is defined as the interface between the electrolyte and the electric conductor. The electrolyte concentration can either be reflected by the potential between the two electrodes or the current flowing between the two electrodes. When the concentration is expressed as a potential difference, the electrochemical sensor is called a potentiometric sensor. When the quantity is expressed as current, it is named an amperometric sensor.

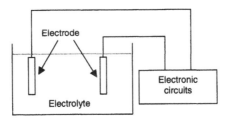

Figure 1.1. An electrochemical measurement system uses electrodes placed in an electrolyte.

At the surface of the metal placed in an electrolyte, there is a potential between the metal and the electrolyte, which is called the electrode potential. A single electrode potential cannot be measured since a potential measurement needs at least two electrodes. If one of them is stable enough to keep a constant potential as a reference electrode, the change of electrode potential of the other electrode can be measured as the difference of the two electrode potentials.

The electrode in an electrolyte reaches equilibrium when no current flows. The electrode potential depends on the concentration (or activity) of the substances involved in the chemical reaction and the electrolyte solution. Electric charge transfers between the boundary of the metal and the solution. Oxidation is a typical reaction for metals.

$$A \leftrightarrow A^+ + ze^-$$ (1.2)

where e^- is an electron, z is the number of electrons involved in the reaction. A is oxidized to A^+ ions and e^- electrons. The metal ions dissolve in the electrolyte. The e^- electrons are left in the metal and make the metal have a negative potential. The potential depends on temperature, the activity of the electrolyte and the activity of metal. It is expressed by the Nernst equation

$$E = E_0 + \frac{kT}{ze} \ln(a_{A^+} / a_A)$$ (1.3)

where E and E_0 are the potentials of the electrolyte and metal respectively, k is Boltzmann's constant (1.38×10^{-23} J K^{-1}), e is the charge of the electron (1.6×10^{-19} C), T is the temperature of the electrolyte, a_{A^+} and a_A are the activities of the metal ion and metal. In a metal electrode, A corresponds to the neutral metal atom bounded in the solid electrode surface, and A^+ corresponds to its ion suspended in the solution. Usually the activity of the ion equals the concentration of the ion in the solution. The activity of the metal is determined by the property of the metal itself.

There are certain kinds of electrode in which the electrode potential is sufficiently stable and is not affected by the change in the concentration of the electrolyte. These electrodes are used as reference electrodes when the potential change is measured. The hydrogen electrode, the silver/silver chloride electrode, and the calomel electrode are typical reference electrodes.

Here we take the silver/silver chloride (Ag/AgCl) electrode as an example. A silver/silver chloride electrode consists of pure silver with a porous layer of silver chloride on its surface. The electrode reaction is

$$Ag \leftrightarrow Ag^+ + e^-.$$ (1.4)

The metallic silver is a pure metal and its activity is unity. So the electrode potential depends only on the concentration of the Ag^+ in the solution. If a solid AgCl and a constant high concentration of Cl$^-$ exist around the electrode, the concentration of Ag^+ is kept almost constant because AgCl partially dissociates into ions as

$$AgCl \leftrightarrow Ag^+ + Cl^-. \tag{1.5}$$

In thermal equilibrium, the product of the concentration of Ag^+ and Cl^- is a constant

$$[Ag^+][Cl^-] = 1.7 \times 10^{-10} \tag{1.6}$$

at 25 °C. When oxidation occurs, the concentration of the Ag^+ is compensated by the AgCl. The above equation remains as long as the concentration of Cl^- is constant and large enough ($[Cl^-]>>[Ag^+]$). Consequentially the electrode potential remains constant.

1.2.2 Potentiometric sensors

Potentiometric sensors measure the potential difference between two electrodes, which reflects the concentration of certain ions. The ion-selective electrode is a typical potentiometric sensor. It consists of two electrodes and one ion-selective membrane between the two membranes (figure 1.2). An ion-selective membrane ideally should be only permeable to a specific ion, which is not practical in real applications. So a membrane highly permeable to one specific ion and less permeable to other ions can be used as an ion-selective membrane.

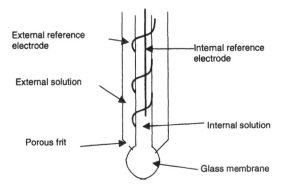

Figure 1.2. The pH electrode uses an ion-selective glass membrane, which is permeable to hydrogen ions.

When immersed in electrolyte, there is ion movement due to the different solution concentrations. However, only the selective ions (say A^+) move into the inner cell through the selective membrane. So the inner solution becomes positive and causes a potential between the internal and external solutions. When the potential is large enough, it forces the A^+ out of the inner cell at the rate of the incoming ions due to the chemical gradient. In equilibrium, the potential between the external solution and internal solution is expressed by the Nernst equation

$$E = E_o - E_i = -\frac{kT}{e}\ln(c_o / c_i) \tag{1.7}$$

where E, E_o and E_i are potential difference, external potential and internal potential respectively. c_o and c_i are the concentration of A^+ in the external and internal solution. If we keep the concentration of A^+ in the internal solution constant (usually using a buffer solution), the potential difference is only dependent on the concentration of the outer solution

$$E = \text{const} - \frac{kT}{e}\ln(c_o). \tag{1.8}$$

The size of the internal tube is much smaller than the volume of electrolyte and it does not affect the concentration of the ions in the electrolyte. Thus the potential is proportional to the logarithm of the ion concentration in the electrolyte.

The most common potentiometric sensor in clinics is the pH electrode with ion-selective glass. Only hydrogen ions (H^+) can pass through the glass and the logarithm of the H^+ concentration is used to reflect the concentration:

$$pH = -\log_{10}[H^+]. \tag{1.9}$$

There are three categories of ion-selective electrodes, the glass electrode, the pressed-pellet ion-selective membrane electrode and the liquid ion-exchange membrane electrode. These membranes are permeable to certain kinds of ion and can be used to measure the concentration of certain ions in solution. Glass electrodes and pressed-pellet electrodes are made of solid materials, such as sodium glass NSA 11-18, solid inorganic materials, single crystals of fluorides and polycrystalline silver sulfide. A liquid ion-exchange membrane is formed by an absorbent material so that ion-exchange material is dissolved in a lipophilic solvent. This type of liquid membrane is not stable in long-term use and needs a polymer-matrix membrane (such as PVC) to extend its life.

There is one kind of potentiometric sensor based on the ion-selective field-effect transistor (ISFET). It consists of a field-effect transistor (FET) and an ion-selective membrane that covers the silicon-dioxide layer on the conductive channel. Figure 1.3 shows an *n*-type transistor on a *p*-type substrate. The current through the *n*-channel from drain to source is controlled by the potential at the gate. However, the insulation layer above the *n*-channel is covered by the ion-selective membrane and exposed to the solution, instead of the metal gate connection. The potential developed by the ion-selective membrane is added to the gate–source voltage, which controls the current in the *n*-channel. With an electronic circuit, the current can be measured and it reflects the ion concentration in the solution.

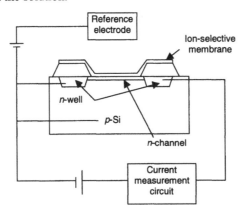

Figure 1.3. The ion-selective field-effect transistor uses an ion-selective membrane to control current in the *n*-channel.

The ISFET provides several advantages compared to a normal ion-selective electrode. The ISFET provides a low-impedance output and does not require the high-impedance amplifier used in glass electrodes. The ion-selective membrane is very thin and provides a faster response time. Fabricated by IC Technology, it is easy to miniaturize and is suitable for mass production. The resulting low cost makes this kind of electrode disposable and avoids the infection issue when used in the body. Currently, there are many companies manufacturing commercial ISFET probes and ISFET pH measurement systems, such as Deltatrack Model 301 (Deltatrack Inc., Pleasanto, CA) and Corning Model 314I (Corning Labware & Equipment, Corning, NY). The ISFET sensor manufactured by DMP Ltd (Hegnau-Volketswil, Switzerland) has an ion-sensitive field-effect transistor with dimensions of 3 mm × 1 mm × 0.3 mm. It features a hydrogen-ion-sensitive surface smaller than 0.4 mm × 0.8 mm. The probe accuracy is 0.01 pH with two point calibration.

1.2.3 Amperometric measurement

An amperometric sensor measures the current drained from the electrode, which reflects the rate of the chemical reaction. If the chemical reaction involves the charge exchange between the electrode surface and the substrate solution, the reaction causes an electric current, which can be measured by an externally connected electronic circuit. The rate of the reaction is governed by the concentration of the solution and the potential applied between the electrode and the solution. If a constant voltage is applied to maintain a certain level, the concentration can be measured by the drained current. Dissolved oxygen and hydrogen peroxide are the common applications that use amperometric measurement.

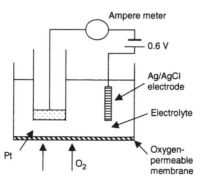

Figure 1.4. The Clark-type oxygen electrode consumes oxygen at the platinum cathode.

Figure 1.4 is a diagram of an oxygen sensor. It contains a platinum cathode and a silver anode. When a voltage of 0.6 V is applied to the electrode, the reaction

$$O_2 + 2H_2O + 4e^- \rightarrow 4OH^- \tag{1.10}$$

occurs at the surface of the platinum electrode. If the amount of oxygen is limited, it is consumed and the reaction stops. But if the oxygen is supplied continuously, the current is constant. An oxygen-permeable membrane covers the electrodes so that the oxygen concentration outside determines the oxygen concentration inside and the rate of the reaction inside.

If oxygen is supplied by diffusion through a membrane of area A and thickness d, the current I in the circuit is approximately

$$I = 4FAD\alpha \, p/d \tag{1.11}$$

where F is Faraday's constant, D is the diffusion coefficient, α is the solubility of oxygen in the solution, and p is the partial pressure of the oxygen. In equation (1.11), the factor 4 is due to fact that four electrons are involved in the reaction with one oxygen molecule. The current is proportional to the oxygen partial pressure, or the oxygen concentration. To maintain the linearity, the geometry of contact, the solubility and diffusion coefficient for oxygen must be kept constant during the process.

When the solution surrounding the electrode does not move, an oxygen concentration gradient builds up due to the oxygen flow into the electrode. Thus the current does not represent the partial pressure in the solution. In a stirred solution, there is no concentration gradient, but the current becomes flow dependent. To avoid this dilemma, a membrane with a lower diffusion coefficient than that of the solution is used to reduce the flow dependence. Also the use of an electrode with a size smaller than the membrane thickness is effective to reduce the flow dependence, because the diffusion field is widely spread at the membrane surface. For the conventional Clark oxygen electrode, the membrane is made of polypropylene or polyethylene and is about 20 μm thick. The distance between the membrane and the electrodes is about 5–10 μm and is filled with electrolyte to maintain the connection.

The detection of hydrogen peroxide is based on the following reaction at the anode:

$$H_2O_2 \rightarrow 2H^+ + O_2 + 2e^-. \qquad (1.12)$$

The reaction is anodic and the working electrode is usually maintained at −0.6 V to provide the bias voltage. Its structure is the same as the oxygen electrode except for the polarity of the applied potential.

1.2.4 Electrochemical gas sensors

Different kinds of gas sensor have been developed based on amperometric and potentiometric methods. The most commonly used are the carbon dioxide (CO_2) electrode based on the potentiometric principle and the oxygen electrode (section 1.2.3) based on the amperometric principle in blood gas measurement.

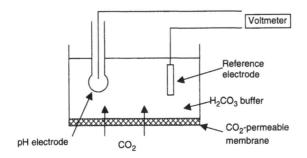

Figure 1.5. In the CO_2 electrode, CO_2 diffuses through the permeable membrane and dissolves in the buffer to form carbonic acid, which is measured by the pH electrode.

The CO_2 electrode (figure 1.5) consists of a pH electrode covered with a gas-permeable membrane such as Teflon and a sodium bicarbonate solution filler between the pH electrode's glass membrane and the gas permeable membrane. This type of electrode is called a Severinghaus electrode. The dissolved carbon dioxide and water forms carbonic acid H_2CO_3. Part of the H_2CO_3 dissociates into H^+ and HCO_3^-. The pH electrode (either glass pH electrode or ISFET) can detect the increased H^+, and thus the concentration of CO_2 from the solution.

Many gas sensors can be built by covering a pH electrode with an appropriate gas-permeable membrane. These sensors, such as CO_2, NO_2, H_2S, SO_2, HF, HCN and NH_3, are available commercially.

In intravascular measurement, the electrochemical sensors are built into catheters, which are about 0.5–1 mm in diameter. Figure 1.6 shows an example of a Clark-type oxygen electrode for intravascular measurement. The catheter tip is housed in PVC to form a 25 μm thick membrane. It has a silver cathode, silver anode and deposited electrolyte at its tip. When it is immersed in blood, the water vapor diffuses through the membrane and forms a liquid electrolyte in about 10–45 min. The sampled blood flows into the catheter through the sample hole and the potential is measured between the two silver wires.

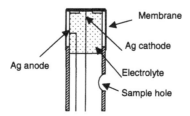

Figure 1.6. In a catheter-type O_2 sensor, blood flows through the sample hole to be measured by a Clark-type oxygen electrode.

An intravascular pCO_2 electrode has also been attempted using a pH electrode surrounded by a CO_2-permeable membrane. But it is more problematic than the pO_2 electrode and has not been widely accepted in clinical gas monitoring.

Also people have tried to combine blood-gas monitoring in one catheter. For example, Parker *et al* (1978) reported a catheter-tip pCO_2 and pO_2 combined electrode. Coon *et al* (1976) reported an intravascular evaluation of a combined pH and pCO_2 electrode. Meruva and Meyerhoff (1998) even reported a combined catheter-type pO_2, pH and pCO_2 electrode.

1.3 FIBER-OPTIC CHEMICAL SENSORS

1.3.1 Spectrophotometric analysis and Beer's Law

The spectrophotometric technique is based on the interaction between optical waves (electromagnetic waves) and molecules, ions and atoms. When an optical wave passes through a medium, it may be absorbed, reflected and scattered by the medium. It may also cause emission, fluorescence and the Raman effect. All these interactions can be explained by quantum physics, which is beyond the scope of this book. They are used to measure different chemical substances in clinical applications. By introducing fiber-optic techniques, optical measurements can even be performed *in vivo*.

In an absorption measurement, a beam of light with a selected wavelength passes through a layer of solution and its attenuation is measured. In a uniform medium, the light is attenuated exponentially according to Beer's Law

$$A = \log \frac{I_0}{I} = aLC \qquad (1.13)$$

where A is the absorbance, I_0 and I are the incident and transmitted light intensity, a is the absorptivity of the sample, L is the length of the optical path and C is the concentration of the absorbing substance. As long as we keep the absorptivity and optical path constant, the light absorbance is proportional to the substance concentration. In real applications, we set up a standard solution with known concentration C_s and absorbance A_s. Then the absorbance A_u of the unknown concentration is determined. The unknown concentration is given by

$$C_u = C_s \frac{A_u}{A_s}. \qquad (1.14)$$

If there are multiple species in the sample, the total absorbance can be expressed as

$$A = L \sum a_i C_i \qquad (1.15)$$

where a_i is the absorptivity for species i, and C_i is the concentration of species i.

For *in vivo* measurement, measurement of light absorption through the substance is not realistic. In such situations, reflected and scattered light intensities are measured to estimate the light absorption of the material. These are very common in fiber-optic measurements.

There are two types of deflection: specular reflection and diffuse reflection. Specular reflection occurs at the interface of two different optical media. Diffuse reflection occurs when the light penetrates into the medium and partly returns to the surface by absorption and scattering.

Scattering is the direction alteration of an optical wave when it is incident on small particles. Rayleigh scattering and Mie scattering are the two common scattering modes in solutions. When the scattering particle is small in comparison with the incident light wavelength, Rayleigh scattering dominates. When the particle is large compared to the wavelength, Mie scattering occurs. The scattered light scatters at a certain angle based on the scattering particles. The scattering light intensity is related to the concentration of the scattering particles.

Molecules, ions and atoms are excited to higher energy levels under some circumstances, such as heat, bombardment of electron and ions, exposure to electromagnetic radiation or chemical substances. When they change back to normal energy levels, they emit light according to the energy-level difference. The emission intensity is related to the chemical concentration of the substance.

Fluorescence is a special emission when atoms, ions and molecules are excited by optical waves. The medium is excited at a short wavelength (high energy) and the fluorescence occurs at a longer wavelength (low energy). Normally the fluorescence is very weak, compared to the excitation light. The measurement system contains an optical filter to filter out the incident light and a sensitive detector to measure the weak fluorescence. At low concentrations, the fluorescence is expressed as

$$I = kI_0abLC \tag{1.16}$$

where I is the fluorescence light intensity, I_0 is the incident light intensity, k is the instrument constant, a is the molar absorptivity, L is the length of optical path, C is the substance concentration and b is the quantum yield of fluorescence. When the lifetime of the fluorescence is measured, it is called phosphorescence.

A spectrophotometer system contains a light radiation source and photodetectors. The light source can be continuous, such as a tungsten filament lamp and a halogen lamp for the visible and near-infrared regions, and a hydrogen lamp for ultraviolet light. To obtain monochromatic light, a prism or a reflection grating can be used. A laser yields a highly monochromatic light, but has a limited number of wavelengths.

The photodetectors are usually photomultiplier tubes and photodiodes. In a photomultiplier tube, an incident photon produces a photoelectron. The emitted photoelectron is amplified by secondary electron generation at the cascaded electrodes. Finally 10^6 to 10^7 electrons are collected at the anode. A photodiode is a semiconductor device in which the current through the reverse biased $p–n$ junction is proportional to the incident light intensity. Fabricated as a linear or two-dimensional array, photodiodes can be used as image sensors too.

1.3.2 Fiber-optic chemical sensors

An optical fiber is an efficient method to transmit light from one point to another. Due to the optical properties of the fiber and its coating, most of the light only transmits down the fiber, instead of escaping out through the surrounding coating. A typical fiber has a diameter of 12–100 μm. A flexible bundle of hundreds of fibers is used in common applications.

Compared with other sensors, fiber-optic chemical sensors have the following advantages.

1.   Optical fibers are thin and flexible. They can form small and lightweight sensors with minimal invasiveness.

2.   Optical fibers do not corrode and can be implanted in the body for a long time without changing properties.
3.   Optical fibers do not interfere with electromagnetic waves. They can eliminate electric and magnetic noise from other equipment. The optical sensor also does not affect other electric devices, such as magnetic resonance imaging (MRI).
4.   Optical fibers use light to transmit signals so the patients are electrically isolated from the electric device. This increases patient safety.
5.   Fiber-optic chemical sensors are an optimum choice where electrodes are not accessible at some sites inside the human body.
6.   The low cost of optical fibers makes the sensor disposable and low priced.

Of course, fiber-optic chemical sensors also have some disadvantages. One main disadvantage is that they sense ambient light, such as surgical lamps. Modulated or coded light can be adopted to avoid this problem.

Fiber-optic chemical sensors measure the changes in optical parameters of the medium with known properties. The optical parameters include absorption, reflectance, scattering, fluorescence and phosphorescence.

A wide range of chemical parameters can be measured with fiber-optic chemical sensors (Harmer and Scheggi 1989). For example, pH, pCO_2, ammonia, phosphate esters and moisture can be measured with absorption methods (absorption, reflection and scattering). Glucose, pH, pO_2, sodium, metal cations and H_2O_2 can be measured with the fluorescence method. The instrument to measure each chemical parameter is designed to match the specific characteristics of the measurement and varies substantially.

In some applications, the fiber is just a passive light conductor. It is not active in the transduction process and only transports photons to and from the medium under study. This type of transducer is defined as an extrinsic sensor. A typical example is the medical probes for oxygen-saturation measurement.

In some other applications, the fiber itself is an active part of the transduction process, such as evanescent mode sensors. The coating layer is usually removed at the end of the fiber, leaving the core exposed to the measurand. The light guided by the fiber has an evanescent wave extending into the surrounding medium and the chemical reaction occurs at a surface layer. This type of fiber-optic sensor is defined as an intrinsic sensor. Evanescent-wave nepholometry is a typical example of this kind of sensor.

The light source for the sensor varies according to the requirement of the measurement. Typical light sources are gas lasers, tungsten lamps combined with filters, light-emitting diodes (LEDs) and semiconductor laser diodes. The photodetectors can be photodiodes, photomultipliers, photoconductive cells and thermal sensitivity cells, depending on the wavelength and intensity of the spectrum.

A pair of optical fibers or fiber bundles can be used for transmitting and receiving the light, respectively. In some cases, both the exciting light and the signal light can transmit on the same fiber with timing switching. In fluorescence, the exciting light and signal light usually have different wavelengths. They can transmit on the same fiber with an optical filter at the receiver end.

Fiber-optic sensors can be used to measure the intravascular blood-gas concentration. Earlier attempts concentrated on the oxygen saturation by reflection spectrometry. Now fluorescence is more common. Figure 1.7 shows a fiber-optic intravascular catheter for pO_2 measurement (Peterson *et al* 1984). The catheter tip has a hydrophobic membrane and inside are fluorescent dye beads such as perylene dibutyrate. The dye is excited with blue light at 486 nm and emits fluorescence at 514 nm. The fluorescence is quenched by oxygen and the effect is dependent on the pO_2. The intensity ratio between the excitation and fluorescence determines the pO_2.

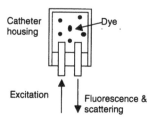

Figure 1.7. In a catheter-type fiber-optic oxygen sensor, oxygen diffuses through the membrane to quench fluorescence in the excited dye.

Other similar techniques have been used to measure chemical concentrations such as CO_2 and H^+. Some catheters have been designed so that the pO_2, pCO_2 and pH can be measured at the same time (Soller 1994).

1.3.3 Optical oximetry

The oxygen saturation of hemochromes, such as hemoglobin, myoglobin and cytochromes, is useful information in evaluation of patients in different clinical situations. Clinical measurements of hemoglobin saturation in major vessels and cardiac chambers are reported during catheterization for detection of cardiac shunts and for monitoring patients with various cardiac diseases. During surgical procedures and recovery from anesthesia, hemoglobin oxygen saturation is utilized for early detection of hypoxia. Oxygen saturation is also useful for the care of fetuses and neonates.

Optical oximetry can be divided into two different types: (1) transmission or forward scattering and (2) reflection or back scattering (figure 1.8). In the forward-scattering mode, a target area is illuminated with a light source. The transmitted or forward-scattered light is detected and analyzed. In the reflection mode, back-scattered light from the specimen is measured to estimate the oxygen saturation. In both methods, absorption and scattering properties of the tissue sample determine the measurement. In particular, wavelength-dependent absorption properties of various species such as hemoglobin, myoglobin, and cytochrome in the tissue affect the measured transmittance and reflectance.

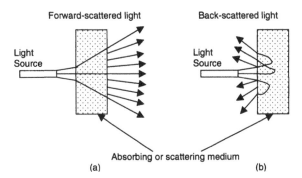

Figure 1.8. Optical oximetry principle. (*a*) Transmission or forward-scattered mode. (*b*) Reflection or back-scattered mode.

In normal tissue, the hemoglobin concentration is much greater than that of other species. When the transmission and reflection spectra are measured, they are strongly dependent on the absorption of hemoglobin. Hemoglobin demonstrates a significantly stronger absorption at the wavelength range between 450 and 550 nm (about 10 times) than the range between 650 and 1000 nm. Bench-top oximeters usually use the spectra around 550 nm to examine the absorption spectrum. The intravascular and tissue oximeters use wavelengths from 650 to 1000 nm. There are two types of hemoglobin in tissue: oxyhemoglobin, which combines with oxygen, and deoxyhemoglobin, which releases oxygen to the tissue and does not have oxygen bound to hemoglobin. At the wavelength of 805 nm, the absorption coefficients of oxyhemoglobin and deoxyhemoglobin are equal and this wavelength is the isosbestic wavelength.

The operation of the oximeter is based on Beer's Law. Consider that two wavelengths λ_1 and λ_2 are used and the concentrations for oxyhemoglobin and deoxyhemoglobin are C_o and C_d. The absorbances for these two wavelengths are

$$\frac{A_1}{L} = a_o(\lambda_1)C_o + a_d(\lambda_1)C_d$$

$$\frac{A_2}{L} = a_o(\lambda_2)C_o + a_d(\lambda_2)C_d.$$

(1.17)

Here $a_o(\lambda_1)$ and $a_o(\lambda_2)$ are the absorptivity of the oxyhemoglobin at wavelength λ_1 and λ_2, respectively. $a_d(\lambda_1)$ and $a_d(\lambda_2)$ are the absorptivity of the deoxyhemoglobin at wavelength λ_1 and λ_2, respectively. From the linear equations above, we can solve the concentrations C_o and C_d. Consequently, the oxygen saturation can be solved as

$$SO_2 = \frac{C_o}{C_o + C_d}. \qquad (1.18)$$

The example above just considers absorbance of the light. When both absorption and scattering are considered, the final absorbance is given as

$$A = B + aCL \qquad (1.19)$$

where B is a term related to the scattering.

During *in vivo* applications, multiple scattering exists in the medium and the absorption length is usually an unknown number. So some empirical equations based on Beer's Law and experimental linear regression are used to estimate the oxygen saturation.

For intravascular oximetry, optical fibers are used to guide the light inside the vessel and to collect the reflected light from blood cells back to the light detector. To estimate SO_2, the reflectance at two wavelengths, one in the red and another in the infrared, are used. The oxygen saturation is based on the empirical relation

$$SO_2 = A + B(R(\lambda_1)/R(\lambda_2)) \qquad (1.20)$$

where A and B are constants determined by the fiber geometry and physiological parameters of blood, and R is the reflectance at the two wavelengths. Currently, intravascular oximeters are used to measure the venous saturation and deduce the status of circulation.

Tissue is a complicated inhomogeneous medium. There are blood vessels, as well as both arteries and veins in the tissue. Red blood cells in arterial blood release oxygen to the tissue and return via the venous system. There are also absorptions due to bone, tissue and skin. Thus analysis in tissue is very complex. Although there were some oximeters designed before the 1980s, they were not fully accepted for clinical use. In the late 1980s, pulse oximeters were developed and became successful to noninvasively measure oxygen saturation in tissue. Now they are standard equipment in operating suites.

Pulse oximetry combines the plethysmographic principle with the optical absorption of hemoglobin (figure 1.9). The absorption in tissue is due to skin pigment, tissue, bone and blood. During the cardiac cycle, the blood flow in veins does not change and the size of the veins remains the same. So the light absorption in skin, tissue, bone and venous blood is constant during the cardiac cycle. On the other hand, there is more blood flowing in arteries during systole than during diastole. The diameter of the arteries increases with pressure. During systole, arteries are largest and cause the largest absorption of light. Hence, the transmitted light has the smallest intensity (I_L). Similarly, the transmitted light is largest during diastole (I_H). Thus the absorption alternates with the pulsation of the arterial blood. The absorption contains two parts. One part is the dc component, which is due to the venous blood, a constant amount of nonpulsating arterial blood, and other nonpulsating parts such as skin, tissue and bone. Another part is the ac component due to the pulsating arterial blood. The alternating part of the light absorption is usually only 1–2% of the dc absorption. This time-varying signal is referred to as the plethysmographic signal.

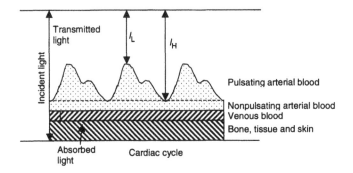

Figure 1.9. Absorbed and transmitted light in blood.

I_H is the light transmitted through the nonpulsating components during diastole. We can assume I_L is the light transmitted through the nonpulsating components plus another layer of artery of thickness ΔL, which is the diameter change of the artery between systole and diastole. Now the incident light intensity on the assumed layer is I_H. So the light transmitted through the layer can be expressed as (consider the oxyhemoglobin and deoxyhemoglobin)

$$A = \log\left(\frac{I_H}{I_L}\right) = (a_o C_o + a_d C_d)\Delta L \tag{1.21}$$

and I_I is given as kI_0, where k is a constant. If we use two lights with different wavelengths, we can compute the oxygen saturation as described before.

Figure 1.10 shows a block diagram for a pulse oximeter. Usually 660 nm (red) is used as one wavelength and 940 nm (infrared) or 805 nm (isosbestic) is used as the second wavelength. The two LEDs alternately emit light through the object. The photodiode receives the transmitted light. The logarithmic amplifier amplifies the signal and feeds it to the infrared (IR) electronic filter and red electronic filter. The analog-to-digital converter digitizes the signal and the microprocessor calculates the oxygen saturation. The results are sent to the display. Also the control signals are sent to the LED driver to adjust the amplitude of the two LEDs.

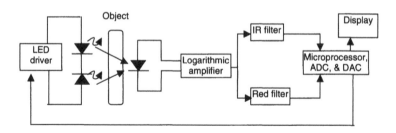

Figure 1.10. In a pulse oximeter, LEDs at 660 nm and 940 nm alternately irradiate a single photodiode.

1.4 OTHER TRANSDUCERS

1.4.1 Acoustic bulk-wave device

The Curie brothers discovered in 1880 that anisotropic crystals, such as quartz, tourmaline and Rochelle salt, give out an electric signal when mechanically stressed. On the other hand, these materials deform mechanically when an electric signal is applied to them. Thus, with an oscillating electric potential, the crystals oscillate mechanically. In addition to crystals, we have ceramic materials such as barium titanate and various lead zirconium titanates and organic polymers, such as polyvinylidene fluoride (PVDF).

Each piezoelectric material has a natural resonant frequency of oscillation, depending on the material and its geometry. Its resonant frequency is also dependent on the mass load on the surface of the resonator. The absorption of an analyte on the surface can cause a frequency change (Δf) of resonance, The change can be measured with a high sensitivity (500–2500 Hz g^{-1}), resulting in a picogram detection limit.

A typical system usually contains two balanced crystals working in differential mode. The surface of the test crystal is exposed to the sample and a chemical reaction takes place there. The reference crystal is isolated from the sample and oscillates at a constant frequency. The frequency difference between these two crystals represents the mass change on the test crystal, hence the chemical concentration in the sample. But it eliminates the effects of other aspects such as temperature and surface size.

An example of the acoustic wave device is the formaldehyde sensor with formaldehyde dehydrogenase (FDH)-NAD$^+$ as the selective layer. The chemical reaction is

$$CH_2O + H_2O + NAD^+ \xrightarrow{\text{FDH}} NADH + HCO_2H + H^+ . \qquad (1.22)$$

1.4.2 Acoustic surface-wave device

In some piezoelectric crystals, such as lithium niobate ($LiNbO_3$), the acoustic wave can travel on the surface of the crystal, instead of inside of the crystal. This type of wave is called a surface acoustic wave (SAW). Various wave modes include (1) Rayleigh surface wave (sometimes also called surface acoustic wave), (2) plate mode wave, (3) evanescent wave, and (4) Lamb wave. The detailed description of these waves is beyond the scope of this book.

SAW devices contain a transmitter and receiver, which are made of interdigitated electrodes. An electric signal applied to the transmitter produces a mechanical wave. This wave travels on the surface of the crystal and produces an electrode voltage at the receiver. The propagation of the surface wave is affected by the load on the crystal surface, which is of the order of several acoustic wavelengths thick. Similar to the bulk-wave device, this change can be used to detect the chemical reaction on the surface, and hence chemical concentration of the sample.

Although much work has been devoted to the development of the SAW chemical device, very few practical applications have reached the market. The greatest difficulty is the thin layer of chemical electrolyte on the surface. Some researchers are investigating antibodies and proteins.

1.4.3 Thermal measurement

There is always heat generation during chemical reactions. The usual amount of heat generated in biochemical reactions is about 25–100 kJ mol^{-1}. The heat elevates the temperature of the solution and can be detected by a thermometric technique (usually thermistor or thermocouple).

The common thermal measurement is a flow-through system to measure the heat production associated with a chemical reaction. The temperature difference between the inlet and outlet is measured to reflect the heat generation, which is relevant to the chemical concentration of the substance involved in the chemical reaction.

1.5 BIOSENSORS

The term 'biosensor' commonly means a device incorporating a biological sensing element either intimately connected to or integrated within a transducer (Turner *et al* 1987). Biosensors generally sense and measure specific chemicals, which need not be biological components themselves. These chemicals are referred to as substrates. Biological elements have a high selectivity to substances so that the measurement can be performed in a mixture with many other substances. With biosensors, the normal separation and purification procedures are not necessary so that *in situ* and *in vivo* measurements can be carried out.

Table 1.1 shows that different biological elements and different transduction principles can be employed to construct a biosensor (Eggins 1996). If the biological element is stable enough, it can be extracted and incorporated in the device directly. If the biological element is not stable when it is extracted from the tissue, there are different approaches in which cells, tissue, or even living organisms can be incorporated in a biosensor. The process of incorporation is called immobilization. Table 1.1 also lists several immobilization methods.

Table 1.1. Biological elements and transduction principles that may be used to construct a biosensor.

Biological elements	Transduction principles	Immobilization methods
Enzymes	Electrochemical	Adsorption
Tissues	Optical	Microencapsulation
Micro-organisms	Thermal	Entrapment
Antibodies	Mechanical (acoustical)	Cross-linking
Nucleic acids		Covalent bonding
Receptors		

1.5.1 Enzyme-based biosensors

An enzyme is a large complex macromolecule, consisting of many proteins, which can catalyze certain chemical reactions. These proteins usually contain a prosthetic group, which often includes one or more metal atoms. Enzyme reactions are highly specific, superior to any synthetic catalysts. An enzyme-based sensor (glucose sensor) was the first biosensor and is still the most common sensor in biosensors.

Enzyme-based biosensors are designed by different transduction methods: electrochemical, optical, thermal and gravimetric. If the electrochemical principle is employed, the enzyme-based sensor is called the enzyme electrode. The enzyme electrode measures either the formation of a product or consumption of a substrate, based on potentiometric or amperometric methods.

The amperometric technique is widely employed in the enzyme electrode. It measures either the consumption of oxygen or formation of hydrogen peroxide. The most important application is the glucose oxidation by the enzymatic reaction of glucose oxidase (GOD)

$$\text{Glucose} + O_2 \xrightarrow{\text{GOD}} \text{Gluoconolactone} + H_2O_2.$$

Here oxygen is the substrate and hydrogen peroxide is the product of the reaction. These both can be measured by the amperometric method (section

1.2.3). This type of glucose electrode was the first and mostly studied biosensor. Figure 1.4 shows a Clark-type glucose electrode. The conventional Clark electrode has an additional layer of enzyme (glucose oxidase), placed on the oxygen-permeable membrane, and a second layer of membrane permeable to glucose (such as cellulose acetate). When the oxygen consumption is measured, the ambient oxygen level should be kept constant since it affects the reaction rate. A reference electrode without GOD may be used to act as a differential electrode to monitor the oxygen-level change. Some chemical methods (such as recycling of H_2O_2 or oxidation of water to oxygen) may regulate the oxygen level. Hydrogen peroxide can be measured by one electrode. Its simplicity makes it advantageous to the oxygen electrode and it is more common than the oxygen measurement.

The above, catalyzed oxidation requires oxygen during the reaction. Fluctuation of oxygen content affects the measurement, especially when the substrate concentration is high. The second generation of biosensor replaced oxygen with other oxidizing agents (mediators) whose concentration is easily controlled. The mediators are electron-transfer agents, which are reversible and have the appropriate oxidation potentials. Most of the mediators are based on iron ions and the most successful mediator is the ferrocenes (Fc). Again, taking glucose as an example, the operation of a mediator is as follows:

$$\text{Glucose} + \text{GOD}_{Ox} \longleftrightarrow \text{Gluconolactone} + \text{GOD}_{Red} + 2H^+$$

$$\text{GOD}_{Red} + 2Fc^+ \longleftrightarrow \text{GOD}_{Ox} + 2Fc$$

$$2Fc - 2e \longleftrightarrow 2Fc^+.$$

Figure 1.11 shows the cyclic process. The glucose is oxidized in the mediator while the Fc is oxidized at the electrode. The current flowing through the electrode is an amperometric measurement of the glucose concentration.

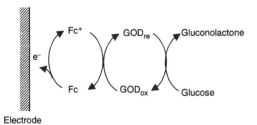

Figure 1.11. In a ferrocene-mediated biosensor for glucose, the glucose oxidase (GOD) enzyme is assisted by ferrocene (Fc).

The third generation of enzyme biosensor was based on direct enzyme–electrode coupling. The enzyme is directly coupled to the electrode, which is made of organic conducting salt (Albery and Craston 1987). The enzyme is 'wired' to the electrode using new immobilization technology.

Many amperometric enzyme electrodes have been developed that can specifically detect different substances, such as glucose, lactate, uric acid, acetylcholine, alcohol, amino acids and so on. Most of them are used in laboratory conditions. But the glucose electrode for the analysis of whole blood is commercially available as a disposable test strip (MediSense, Inc., Cambridge, MA). It measures the glucose level just from one drop of blood from the finger.

The enzyme electrode can also be built based on potentiometric measurement. A typical example is the hydrolysis of urea catalysed by urease as

$$CO(NH_2)_2 + H_2O \xrightarrow{\text{urease}} CO_2 + 2NH_3 .$$

The concentration of ammonia is monitored as a voltage by an ammonium ion-selective electrode. This voltage is proportional to the logarithm of the concentration of ammonia, which is directly proportional to the concentration of urea. The ammonium ion-selective electrode is really a modified pH electrode (section 1.2.1). The enzyme is mixed with a gel and coated on a nylon net membrane covering the electrode. It is held in place with another (dialysis) membrane.

1.5.2 Immunosensors

The immunosensor utilizes an antigen–antibody reaction to measure the chemical concentration, similar to the analytical technique called immunoassay. The binding between antigen and antibody is more powerful and specific than the binding between enzymes and their substrate. Thus immunosensors are more sensitive than enzyme sensors.

The antigen A_g and antibody A_b form the antigen–antibody complex (A_gA_b) as

$$A_b + A_g \longleftrightarrow A_bA_g .$$

The reaction is reversible and its affinity constant, $K = [A_bA_g]/[A_b][A_g]$, is usually about 10^6. When A_g is introduced and the amount of A_b is kept constant, the introduced A_g can be measured by the production of A_bA_g. Usually the antibody is immobilized and fixed on the surface of the sensor, and the formation of A_bA_g causes some changes of electric potential or mass on the surface of the sensor. Some research groups used the immunochemically sensitive FET (IMFET) or SAW device to measure these changes (Janata and Blackburn 1984, Roederer and Bastiaans 1983).

The immunosensor can also be designed with an enzyme reaction, in which a fixed amount of enzyme-labeled antigen is introduced. The unlabeled and labeled antigen can be bound to the immobilized antibody on the surface of the sensor. Then the free antigens are removed and the activity of enzyme is measured by introducing another substrate. The formation of the product can be measured by either electrochemical or optical methods. The binding reactions of the labeled and unlabeled antigen are competitive. When the concentration of unlabeled antigen is high, there are more antibodies occupied by the unlabeled antigens and fewer by labeled antigens. Consequently, after removing the free antigen, the measured enzyme activity decreases. Due to the catalytic function of enzymes, one molecule of enzyme can produce a great number of product molecules. This results in extremely high sensitivity.

Due to their high sensitivity, immunosensors only need very small amounts of sample and detect very low chemical concentrations from the sample. They can be built to a very small size by using semiconductor technology. There is very active research on immunosensors. The Naval Research Laboratory (Washington, DC) and LifePoint Inc. (Rancho Cucamonga, CA) have developed instruments using immunosensors to measure different small molecules such as explosives, pollutants, hormones and therapeutic drugs.

1.5.3 Microbial sensors

Microbial biosensors incorporate a micro-organism-sensing element that specifically recognizes the species of interest. Like enzyme-based biosensors and immunosensors, the sensing element is either intimately connected to, or integrated within, a suitable transducer system. Microbial sensors have selectivity similar to enzyme-based sensors, but there are some advantages of the micro-organism element over the isolated enzyme:

1. They are less sensitive to inhibition by solutes and are more tolerant of suboptimal pH and temperature values than enzyme electrodes.
2. They have a longer lifetime.
3. They do not need isolation as enzyme electrodes do and are less expensive.

However, microbial sensors also have disadvantages such as:

1. Some have a longer response time than the enzyme electrode.
2. They need more time to return to the base line after use.
3. Micro-organisms contain many enzymes and care must be taken to ensure selectivity.

Microbial sensors convert biochemical signals to electric signals. They usually consist of a membrane with immobilized micro-organisms and an electrochemical device. Their sensing mechanism is classified as respiration-activity measuring or metabolite-activity measuring.

In the case of respiration-activity measurement, changes (normally increases) in respiration of micro-organisms caused by assimilation are detected by an oxygen electrode. The sensor consists of immobilized aerobic micro-organisms and a Clark oxygen electrode. The sensor is put in a buffer saturated with oxygen. With the addition of the substrate, the metabolic reaction occurs and the consumption of oxygen (decrease of oxygen) is measured to calculate the concentration of the substrate. Glucose, assimilatable sugars, acetic acid, ammonia and alcohol can be measured by the respiration method.

A metabolite-activity sensor consists of immobilized micro-organisms and a sensor that detects the metabolite produced by the reaction catalyzed by the micro-organism. Using different types of gas and ion detector, many kinds of substance can be detected by different metabolites. For example, the CO_2 electrode is used for measuring glutamic acid or lysine. The pH electrode can be used for measuring cephalosporin and nicotinic acid.

PROBLEMS

1.1   Calculate the pH for a hydrogen ion concentration of 0.01 mol l^{-1}.

1.2   Explain how to avoid the high-impedance amplifier required for the glass pH electrode.

1.3   Give the chemical reaction that occurs at each electrode of an oxygen electrode.

1.4   Give the chemical reaction that forms the basis of the CO_2 electrode.

1.5   For a spectrometer, a 10 g l^{-1} concentration standard yields $I_0/I = 0.3$. An unknown yields $I_0/I = 0.2$. Calculate the concentration of the unknown.

1.6   Describe the operation of a fiber-optic chemical sensor.

1.7   Explain how pulse oximetry differs from regular oximetry.

1.8   Explain how piezoelectric materials are used to sense chemicals.

1.9   Sketch the construction and circuit for an enzyme glucose sensor.

1.10   Explain the operation of an immunosensor.

REFERENCES

Albery W J and Craston D H 1987 Amperometric enzyme electrodes: theory and experiment *Biosensors: Fundamentals and Applications* ed A P F Turner, I Karube and G S Wilson (London: Oxford University Press)

Coon R L, Lai N C J and Kampine J P 1976 Evaluation of a dual-function pH and pCO_2 *in vivo* sensor *J. Appl. Physiol.* **40** 625

Eggins B 1996 *Biosensors: An Introduction* (Stuttgart: Chichester and Teubner)

Harmer A and Scheggi A 1989 Chemical, biochemical, and medical sensors *Optical Fiber Sensor Systems and Applications* ed B Culshaw and J Dakin (Norwood, MA: Artech House)

Janata J and Blackburn G G 1984 Immunochemical potentiometric sensors *Ann. N.Y. Acad. Sci.* **428** 286

Meruva R K and Meyerhoff M E 1998 Catheter-type sensor for potentiometric monitoring of oxygen, pH and carbon dioxide *Biosensors Bioelectron.* **13** (2) 201-12

Parker D, Delpy D and Lewis M 1978 Catheter-tip electrode for continuous measurement of pO_2 and pCO_2 *Med. Biol. Eng. Comput.* **16** 599

Peterson J I, Fitzgerald R V and Buckhold D K 1984 Fiber-optic probe for in vivo measurement of oxygen partial pressure *Anal. Chem.* **56** 62–7

Roederer J and Bastiaans G 1983 Microgravimetric immunosassay with piezoelectric crystals *Anal. Chem.* **55** 2333

Soller B R 1994 Design of intravascular fiber optic blood gas sensors *IEEE Eng. Med. Biol.* **13** (3) 327–35

Takatani S and Ling J 1994 Optical oximetry sensors for whole blood and tissue *IEEE Eng. Med. Biol.* **13** (3) 347–57

Turner A P F, Karube I and Wilson G S (ed) 1987 *Biosensors: Fundamentals and Applications* (London: Oxford Science)

Webster J G (ed) 1997 *Design of Pulse Oximeters* (Bristol: Institute of Physics)

CHAPTER 2

NEURO-ELECTRIC SIGNAL RECORDING

Hong Cao

2.1 NEURO-ELECTRIC SIGNAL

2.1.1 Resting potential

Bioelectric potentials represent the electrochemical activity of a certain class of cells, called excitable cells. These cells are the basic components of nervous, muscular or glandular tissue. The excitable cell exhibits a resting potential in steady state and an action potential when stimulated.

The individual cell maintains a steady electric potential difference between its internal and external environments. This resting potential is normally between −50 and −100 mV, relative to the external cell fluid. It is produced as a result of the unequal concentrations of sodium (Na^+) and potassium (K^+) ions between the internal cell fluid and external cell fluid.

The cell membrane is a very thin (7–15 nm) lipoprotein layer. It is essentially impermeable to intracellular proteins and other organic ions. In the resting state, it is moderately permeable to Na^+ and rather freely permeable to K^+ and Cl^-. The permeability of the membrane for K^+ is about 50 to 100 times higher than that for Na^+. The ion concentrations of ions are severely imbalanced between the internal cellular fluid and external cellular fluid.

Normally the K^+ concentration of the intracellular fluid (~150 mmol l^{-1}) is much higher than that of the extracellular fluid (~3 mmol l^{-1}). Thus there is a K^+ diffusion gradient across the cell membrane and K^+ ions flow from the inside to the outside of the cell. The leakage of positive ions makes the inside of the cell negative and the outside of the cell positive. A transmembrane potential is established and it tends to inhibit the K^+ outflow. The diffusion and electric forces acting across the membrane are opposed to one another and a steady state is ultimately achieved. The membrane potential at this steady state is called the equilibrium potential for potassium E_K. It can be calculated from the Nernst equation

$$E_K = \frac{kT}{ze} \ln \frac{[K^+]_o}{[K^+]_i} = 0.0615 \log_{10} \frac{[K^+]_o}{[K^+]_i} \quad V .$$

Here $[K^+]_o$ and $[K^+]_i$ are the extracellular and intracellular concentration of K^+. A normal E_K value is about -100 mV.

Also the Na^+ concentration of the intracellular fluid ($\sim$10 mmol l^{-1}) is much smaller than that of the extracellular fluid ($\sim$140 mmol l^{-1}). The equilibrium potential for sodium E_{Na} is about 70 mV. There exist both Na^+ and K^+ in the cell, but the final equilibrium potential is not just a simple sum of E_K and E_{Na}. As the permeability of K^+ is much higher than that of Na^+, the K^+ ion flow dominates the ion exchange of the cell membrane. So the equilibrium potential for the cell is close to E_K and is negative. Considering the effect of the Na^+ flow, the total equilibrium is raised a little bit and is usually about -70 mV. Sherwood (2001) has a detailed description of the cell membrane potential.

2.1.2 Action potential

The cell membrane is said to be polarized because of the negative resting potential. When a stimulus brings a depolarization in the membrane and increases the potential to exceed the threshold, it elicits an all-or-none action potential (figure 2.1). The all-or-none property means that the membrane potential goes through a characteristic cycle: a change in potential from resting potential to a certain amount for a fixed duration. For a nerve fiber, the change is about 120 mV and the duration is approximate 1 ms. Further increase of intensity or duration of the stimulus produces the same results.

The origin of the action potential lies in the potential and time-variant dependence of the membrane permeability to the Na^+ and K^+ ions. As the membrane potential increases and exceeds the threshold, the sodium channels in the membrane open and the Na^+ permeability significantly increases. Large numbers of positive Na^+ ions rush into the cell and cause depolarization of the membrane potential. After about 1 ms, these additional sodium channels close and the membrane potential stops increasing at a certain potential level. Meanwhile, the potassium channels also open but at a slower pace and longer duration ($\sim$ 3 ms). After the sodium channels close, the potassium channels are still open and cause repolarization. The membrane potential decreases and returns back to the normal resting potential after an undershoot due to the hyperpolarization. Sherwood (2001) has a detailed description of the action potential.

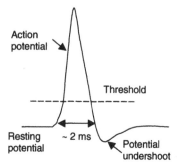

Figure 2.1. Action potential. When the membrane potential exceeds the threshold, it elicits a characteristic potential rise called an action potential.

The action potential is the basis for transmitting a nerve stimulus. If a stimulus propagates by the charge it stimulates, the electric signal decreases as a result of the resistance of the tissue and cannot transmit more than several centimetres. But when a stimulus causes an action potential in a cell, the generated action potential can stimulate the cells in the vicinity and generate action potentials in other cells with the same amplitude and duration. As this process goes on, the stimulus regenerates itself during propagation and can transmit along the nerve, even more than a metre.

2.2 CONVENTIONAL ELECTRODES

From a group of tissues or even organs we can measure the bioelectric potential, such as the electromyogram (EMG), electrocardiogram (ECG), electroretinogram and electroencephalogram (EEG). These biopotentials can be recorded with surface electrodes that contact the skin or internal electrodes that are inserted into the body. Neuman (1998) provides a detailed description and analysis of these electrodes.

The recording of biopotentials generated by individual nerve cells in the central and peripheral nervous system is important in brain function studies. It helps us to understand the basic structure of neurons and the neural system, the interaction among neurons and how the brain processes information to control all basic bodily functions. Also the recording and processing of control signals from neurons is essential for closed-loop neural prostheses.

When a neuron receives sufficient stimuli from other cells, its cell membrane depolarizes, causing ionic current to flow in the extracellular fluid. The voltage change associated with this extracellular current can be sensed with a suitable probe placed in the vicinity of the neuron. Typically, an extracellular signal has amplitude of 50–500 µV and frequency content from 100 Hz to 6 kHz. To record these action potentials, the electrode must penetrate the narrow clefts between cells and approach the active neuron without irreversibly damaging the neuron and stimulating cells. The neurons are usually 50 µm or less in diameter. Here the most important issue is the size of the probe, which should be smaller or at least comparable to the size of neurons. In practice, the probe should be as small as possible and have a shape that facilitates the entry and movement through tissue.

2.2.1 Metal microelectrode

The metal microelectrode is essentially a fine needle of strong metal with insulation up to its tip. The needle is etched as it is slowly withdrawn from an electrolyte solution. By varying the composition of the etching solution and the speed with which the needle is withdrawn from the solution, we can control the shape and size of the needle tip. The electrode is usually made of platinum, gold, stainless steel, tungsten or molybdenum. The tip diameter of the finished needle can be quite small (down to 0.1 µm). The etched metal electrode is then attached to a larger metal support shaft. The electrode and the metal shaft are insulated with a film. Only the tip of the electrode remains uninsulated.

Due to the electrical properties of the metal electrode, it behaves as a high-pass filter to neuro-electric signals (Neuman 1998). The metal microelectrode is suitable for recording neural signals with a frequency range from 100 Hz to 6 kHz.

2.2.2 Micropipette electrode

The glass micropipette electrode is made from 1 to 2 mm diameter glass capillaries. The glass tube is heated up to the softening temperature and then rapidly stretched into halves to make fine tips at the breaking point. By controlling the temperature and stretching force, the wall thickness and shank taper can be satisfactorily manipulated. The resulting tip usually has a diameter of the order of 1 µm, and a tip size of 0.1 µm has also been reported. After the tip is made, the micropipette is filled with electrolyte (usually 3 M KCl). A metal electrode (Ag/AgCl) is introduced into the pipette. The electrode contacts the neuron through the fluid junction at the tip.

Unlike the metal microelectrode, the junction potentials between the pipette electrolyte and intracellular electrolyte and the potential of the Ag/AgCl electrode keep constant during measurement. So the micropipette electrode is suitable for the measurement of dc and low-frequency signals. However, due to its structure, it behaves as a low-pass filter and its bandwidth is limited to a few kilohertz (Neuman 1998).

2.3 SILICON-BASED MICROELECTRODES

The photo-engraved microelectrodes are fabricated with the same technology used in the silicon integrated circuit industry. The thin-film electrode is deposited and patterned on a thick substrate. The top and bottom of the thin-film electrode are insulated with dielectric materials such as silicon oxide, polymethylmethacrylate (PMMA), polyimide, silicon nitride and glass. The recording sites are then patterned and etched through the top dielectric material. The finished electrode is removed from the host substrate. The substrate could be silicon, tungsten, molybdenum, glass, polyimide or other insulator. But silicon is the most used substrate. The electrode conductor materials include gold, platinum, tungsten, tantalum and nickel.

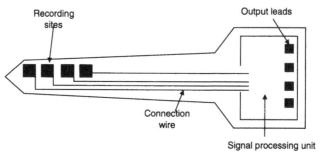

Figure 2.2. A multichannel silicon recording microprobe uses metal electrodes at recording sites (revised from Najafi 1994).

Figure 2.2 (Najafi 1994) shows the basic structure of the silicon microelectrode. The probe consists of a precisely micromachined silicon substrate. The recording sites are patterned according to the design. With the microelectromechanical systems (MEMS) technology, the size of recording sites and their distances are well controlled to the order of micrometres. The number of recording sites and pattern of the recording sites can be varied according to the requirements. This is of great advantage compared to the conventional metal microelectrode and glass micropipette electrode.

After removal from the host substrate, the silicon substrate of the electrode is boron-doped silicon substrate, which is about 8–15 μm thick. This substrate is capable of being elastically bent and it is reported that it can bend as great as 90°. The typical shank length is about 1.5–3 mm. The shank width is normally tapered from 10 μm at the tip to 50 μm near the end. When implanted, the output leads are connected to the outside equipment using newly developed silicon-based multilead ribbon cables.

When the electrode contains only recording sites and connection wires, it is called a passive electrode. When the probe contains electronic circuitry on the same substrate, it is defined as an active electrode. Although only a few attempts have been tried so far to fabricate active electrodes, it is still desirable in clinical use. The on-chip circuits can potentially minimize the output leads and bonds, and reduce the effects of the leakage from the output wires. In addition, on-chip circuitry can be used to amplify the recorded signal and minimize the effect of stray capacitance. It is also essential for prosthetic applications, where the probes are interfaced with computers to sense or stimulate the nerve. Najafi and Wise (1986) reported an active recording array with an on-chip signal processing unit. The active probe is still at its initial stage of development.

In addition to recording neural signals in two dimensions (i.e. depth and lateral), the electrodes can be formed as an array to record neural signals in three dimensions. Hoogerwerf and Wise (1991) and Campbell *et al* (1991) both developed these. In Hoogerwerf and Wise's design, several 32-electrode active multishank probes are precisely positioned on a micromachined platform. The probes are inserted through the slots in the platform. Some spacebars were used to keep them parallel during assembly.

Another application of silicon-based microelectrode is the nerve regeneration electrode used to record and stimulate neural signals from the peripheral nervous system (Blau *et al* 1997, Najafi 1994). The thin silicon diaphragm has many (many more than illustrated in figure 2.3) small holes, which have a diameter of 5 μm. The diaphragm is positioned between the cut ends of a peripheral nerve. The nerves regenerate through the holes and reinnervate on the target tissue or organ. Some of the holes have connection wires to act as recording sites. As the positions of the recording sites and nerve are fixed, it can be used for long-term recording and monitoring.

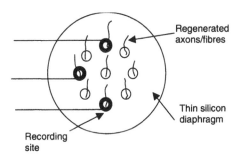

Figure 2.3. Diagram for sieve electrode. The silicon diaphragm has holes and nerves can regenerate through these holes to the target organ. Some of the holes have recording sites for recording neural signals. These recording sites can also be used as stimulating sites.

PROBLEMS

2.1 Calculate the resting potential for $[K^+]_i = 155$ mmol l^{-1} and $[K^+]_o = 5$ mmol l^{-1}.

2.2   List advantages and disadvantages of metal and glass microelectrodes.

2.3   Describe the fabrication of a silicon multielectrode microprobe.

2.4   Explain how a sieve electrode could be used in an artificial arm.

REFERENCES

Blau A, Ziegler Ch, Heyer M, Endres F, Schwitzgebel G, Matthies T, Stieglitz T, Meyer J U and Gopel W 1997 Characterization and optimization of microelectrode arrays for *in vivo* nerve signal recording and stimulation *Biosensors Bioelectron.* **12** 883–92

Campbell P K, Jones K E, Huber R J, Horch K W and Normann R A 1991 A silicon-based three-dimensional neural interface: manufacturing processes for an intracortical electrode array *IEEE Trans. Biomed. Eng.* **38** 758–67

Hoogerwerf A C and Wise K D 1991 A three-dimensional neural recording array *Dig. Int. Conf. Solid-State Sensors Actuators* 120–23

Najafi K 1994 Solid-state microsensors for cortical nerve recordings *IEEE Eng. Med. Biol.* **13** (3) 375–87

Najafi K and Wise K D 1986 An implantable multielectrode recording array with on-chip signal processing *IEEE J. Solid-State Circuits* **21** 1035–44

Neuman M R 1998 Biopotential electrodes *Medical Instrumentation: Application and Design* 3rd edn ed J G Webster (New York: John Wiley)

Sherwood L 2001 *Human Physiology* 4th edn (Pacific Grove CA: Brooks/Cole)

CHAPTER 3

PRESSURE SENSORS

Hong Cao

3.1 PRESSURE MEASUREMENT

Pressure is the force exerted per unit area. Its unit is pascal or Pa (N m^{-2}). In clinical applications, the unit of millimeters of mercury (mmHg) or centimeters of water (cmH$_2$O) is more often used. The conversion of these units is

$$1\,\text{mmHg} = 133.3\,\text{Pa}$$
$$1\text{cmH}_2\text{O} = 98.1\,\text{Pa}.$$

Physiological pressure usually uses atmospheric pressure as a reference pressure. It is measured and expressed relative to the atmospheric pressure. It is not necessary to specify the absolute value of the pressure. However, with some types of transducers, especially implantable pressure sensors, the measured pressure is the absolute value.

Pressure in the body is measured as part of the clinical examination. The most common pressure measurements are the pressures in the cardiovascular system, which represent the performance of the circulatory system. The most important pressure measurements are the arterial blood pressure, mean arterial pressure, left ventricular pressure, right ventricular and pulmonary pressures and central venous pressure. Table 3.1 lists the locations and purposes of these pressure measurements.

Besides cardiovascular pressure, other pressures are measured in different cavities and conduits inside the body. Table 3.2 lists some common pressure measurements and their normal range.

33

Table 3.1. Cardiovascular pressure measurements and their purposes.

Location	Purpose
Arterial pressure	As an index of the circulatory condition
Mean arterial pressure	Characteristics of the whole circulatory system
Left ventricular pressure	Pumping action of left ventricle to the circulatory system
	Assessment of the cardiovascular function
Right ventricular and pulmonary pressure	Diagnosis for pulmonary arterial disease or ventricular septal defect
Central venous pressure	Index of blood volume in the venous system and the elasticity of the veins
	Index of the cardiac function

Table 3.2. Pressure measurements in the body.

	Location	Normal range
Intracranial pressure	Cerebrospinal space, ventricles of the brains	< 8 mmHg
Intraocular pressure	Corneal surface	10–20 mmHg
Intrauterine pressure	Amniotic cavity	40–80 mmHg
Urinary bladder pressure	Bladder	Urinary reflex causes 50 cmH$_2$O
Intrapleural pressure & intratracheal pressure	Lung	5 cmH$_2$O
Intragastric pressure & intraintestinal pressure	Alimentary canal	30 mmHg

3.2 INDIRECT PRESSURE MEASUREMENT

The ideal noninvasive way to measure the pressure is from outside the body (indirect pressure measurement). However, pressures that can be measured indirectly are limited. The most successful clinical application of indirect pressure measurement is the arterial blood pressure using the occlusive cuff technique. It has been widely used and is a standard routine in clinical diagnosis.

A typical sphygmomanometer consists of an inflatable cuff for occlusion of blood, a rubber bulb for inflation of the cuff and either a mercury or aneroid manometer for measurement of pressure. First the occlusive cuff is inflated until the cuff pressure is higher than the systolic pressure. Then the cuff is deflated at 2–3 mmHg s^{-1}. When the cuff pressure is between the systolic pressure and diastolic pressure, the blood under the cuff spurts and causes a palpable pulse in the wrist. The vibration of the vessel under the cuff generates an audible sound (*korotkoff* sound), which can be heard by a stethoscope. At the first audible sound, the pressure read from the manometer indicates the systolic pressure. The last audible sound indicates the diastolic pressure (Peura 1998).

The more advanced automated blood measurement system contains an electric motor and other devices to inflate and deflate the cuff automatically. The palpable sound is detected with a microphone or ultrasonic transducer. Some processors are used to analyze the signal and calculate the blood pressure and heart beat. A typical unit currently costs about $50–$100.

Another indirect pressure measurement is tonometry. When a pressurized vessel is partially collapsed by an external object, the circumferential stress in the vessel wall is relieved. Under that circumstance, the internal pressure is equal to the external pressure exerted. This method is used successfully to measure the intraocular pressure. In the clinical standard procedure, the optometrist applies a force to the corneal surface with a probe. The cornea is flattened out with the advance of the probe. The optometrist measures the force required to flatten a specific optically determined area with a force transducer. That force balances the internal pressure of the cornea and can be used to calculate the intraocular pressure. This procedure is called applanation tonometry.

Nowadays, a noncontact optical applanation tomometer is more often used (figure 3.1). The air valve sends out a linearly increasing air pulse and flattens out the cornea. The light emitter transmits a collimated beam of light at the corneal vertex and a light detector receives the light reflected from the cornea. When the cornea is undisturbed, the curvature reflects a collimated beam at different angles and little light is received by the detector. As the cornea's convexity is progressively reduced by the air pulse, the amount of light received by the detector increases accordingly. When the cornea is applanated, it acts like a planar mirror and reflects the whole collimated beam to the detector. Consequently the detector achieves maximal signal and the air pulse force then is the intraocular pressure. When the cornea becomes concave, a sharp reduction of the received light occurs. At that time, the air pulse should be shut off to avoid further impinging the cornea.

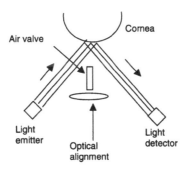

Figure 3.1. In noncontact applanation tomometry, an air puff flattens the cornea and maximizes light detection.

3.3 DIRECT MEASUREMENT

3.3.1 Diaphragm for pressure sensor

Most pressure sensors for direct measurement have an elastic diaphragm. When a pressure difference on the two sides of the diaphragm exists, it causes a deformation of the diaphragm. Although the deformation is nonlinear, it is regarded as linear when the diaphragm is thin compared to the planar size of the diaphragm and the deformation is small compared to the thickness. A simple example is the deformation of a circular diaphragm with clamped edge (figure 3.2). The displacement in the diaphragm (Togawa *et al* 1997) is

$$z(r) = \frac{3(1-\mu^2)(R^2 - r^2)^2 \Delta P}{16Et^3} \qquad (3.1)$$

where μ is Poisson's ratio, E is Young's modulus, R is the radius of the diaphragm, t is the thickness, and ΔP is the pressure difference. At the center of the diaphragm, the displacement $z(0)$ is

$$z(0) = \frac{3(1-\mu^2)R^4 \Delta P}{16Et^3}. \qquad (3.2)$$

The deformation induces both radial strain and tangential strain in the diaphragm. These strain components are equal at the center, and given by

$$\varepsilon = \frac{3(1-\mu^2)R^2 \Delta P}{8t^2 E}. \qquad (3.3)$$

Also the volume change due to the deformation is

$$V = \frac{\pi(1-\mu^2)R^6 \Delta P}{16Et^3}. \qquad (3.4)$$

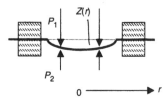

Figure 3.2. Deformation of a thin circular diaphragm with clamped edge. $Z(r)$ is the deformation.

Equations (3.1)-(3.4) provide a simplified view of the characteristics of the diaphragm sensors. The center displacement and strain are proportional to the pressure difference between the two sides. For a given pressure, its response is determined by the geometry of the diaphragm and its material properties. For minimal invasiveness, we want to reduce the size of the diaphragm. As long as we keep the ratio of R/t constant, diaphragm sensors of different size have the same strain and relative displacement ($z(0)/R$). In other words, these sensors with similar geometry and material properties have the same sensitivity no matter the size of the sensor. For the same sensitivity, it is apparent that a smaller sensor is advantageous since it causes less pain when inserted into the body.

Very small diaphragms can be made by microelectromechanical systems (MEMS) technology. Based on matured silicon technology used to manufacture integrated circuits (IC), the sensor can be made with an accuracy on a micrometer scale and has a size of several hundred micrometers (Tohyama *et al* 1998, Wu *et al* 1993). The lower limit of the size is determined by the noise level due to the Brownian motion of the molecules, which is insignificant for physiological pressure measurement.

3.3.2 Strain-gage pressure sensor

A strain gage is based on the geometrical change and variation of resistance of a conductor or semiconductor when subject to a mechanical deformation. For a wire with a length L, cross section A and resistivity ρ, the electric resistance R is

$$R = \rho \frac{L}{A}. \tag{3.5}$$

Under deformation, the length, ΔL, cross section ΔA, and resistivity $\Delta \rho$, may change and may cause change of the resistance ΔR,

$$\frac{\Delta R}{R} = (1 + 2\mu)\frac{\Delta L}{L} + \frac{\Delta \rho}{\rho}. \tag{3.6}$$

Dimensional   Piezoresistive
effect   effect

The change in resistance consists of two parts: a dimensional part $(1+2\mu)\Delta L/L$, which is due to the change in length and cross-sectional area, and a resistive part $\Delta \rho / \rho$, which is due to the strain-induced resistivity change in the structure of the material (piezoresistive effect).

The gage factor (G) is used to compare different strain-gage materials:

$$G = \frac{\Delta R/R}{\Delta L/L} = (1 + 2\mu) + \frac{\Delta \rho / \rho}{\Delta L / L}. \tag{3.7}$$

A larger gage factor means a larger resistance change for a given deformation. The gage factor for metals is mainly due to the dimensional change, and that of semiconductor materials is dominated by the piezoresistive effect. The gage factor of semiconductor materials such as silicon and germanium is approximately 50–70 times that of metals. For example, *p*-type silicon has a gage factor of 100–170 and *p*-type germanium 102. However, the semiconductor materials also have a bigger dependence on the temperature change and should have temperature compensation.

A strain gage is usually connected as a Wheatstone bridge in applications to reduce the temperature effect (Pallás-Areny and Webster 2001). An excitation voltage is applied across two opposite ports of the bridge and the output voltage is measured between two other ports. The output changes according to the change in the resistors under deformation.

Figure 3.3 is an example of a pressure sensor based on a silicon strain gage (Wu *et al* 1993). The silicon substrate was an *n*-type silicon wafer. The four piezoresistive arms were formed by implanted boron ions and have a size of 180 μm × 320 μm. The cavity is etched with catalysed ethylenediamine–pyrocatechol–water (EPW) solution. The thickness of the diaphragm is 15 μm. After connecting all wires and packaging, the chip size is 1.0 mm × 2.5 mm. The sensitivity is about 100 μV (V kPa)$^{-1}$ and nonlinearity is less than 0.2% of the full scale.

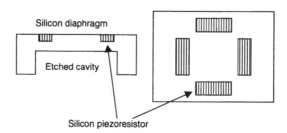

Figure 3.3. Pressure bends the silicon diaphragm in a semiconductor strain-gage type pressure sensor (a simplified model from Wu *et al* 1993).

With the advance of MEMS technology, the strain gage can be designed on the silicon wafer in different shapes, such as plate type or beam type. It also eliminates the mechanical or thermal instabilities caused by the bonding of the strain gage on an elastic material. The silicon sensor can be manufactured with ordinary integrated-circuit processing technology. It is very suitable for mass production and its low cost makes it disposable. Many strain-gage pressure sensors are available, such as P1OEZ (Spectromed, Oxford, CA) and CDXIII (Code Lab., Inc., Lakewood, CA). They both have a range from –30 to 300 mmHg.

3.3.3 Capacitive pressure sensor

If two plate electrodes are arranged parallel and close enough so that the fringe effect can be ignored, the capacitance between these two plates is

$$C = \frac{\varepsilon A}{d} \qquad (3.8)$$

where ε is the dielectric constant of the medium (8.85×10^{-12} F m^{-1} in air), A is the area, and d is the separation distance. If one electrode is fixed to the substrate (or housing), and another electrode is attached to the diaphragm, the distance changes with the diaphragm. So the pressure on the diaphragm can be detected by the measurement of the capacitance (as shown in figure 3.4(a)).

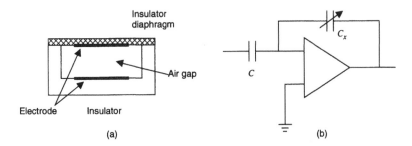

Figure 3.4. Pressure on the diaphragm of a capacitive pressure sensor changes the capacitance (a) and a circuit yields linear output with the change of the displacement (b).

For the capacitance pressure sensor, there are no mechanical contact, friction or hysteresis errors in the measurement. Also it is highly stable and reproducible. It has minimum dependence on the temperature because the dielectric constant changes little with temperature. With the advance of MEMS technology, the circuits needed to make the measurement can be manufactured on the same wafer, very close to the sensor. Thus the interference of stray capacitance can be reduced to a minimum and the sensor still has a very small size.

A problem with the capacitive pressure sensor is that the capacitance change is not linear with the distance. But a linear output still can be achieved by some methods.

Instead of capacitance, we can measure the impedance, which is $1/j\omega C = d/j\omega\varepsilon A$, as a linear function of displacement. Another method is to build a circuit as shown in figure 3.4(b). C_x is the pressure sensor and C is a known capacitance. When a sinusoidal wave $V_i e^{j\omega t}$ is applied to the input of the amplifier, the output of the amplifier is

$$V_o = \frac{d}{\varepsilon A} C V_i e^{j\omega t} \tag{3.9}$$

which is linear with displacement.

A differential capacitor also gives an output linearly dependent on the displacement. Figure 3.5 shows a typical configuration of a differential capacitor. The upper and lower capacitors C_1 and C_2 initially have the same dimension. The center diaphragm is where pressure is exerted and changes the displacement x. When a sinusoidal voltage V is applied and we measure the difference between V_1 and V_2 (Pallás-Areny and Webster 2001),

$$V_1 - V_2 = V\frac{x}{d} \tag{3.10}$$

which is linear with the displacement of x. Other configurations, such as change of overlapped area, give a capacitance proportional to the displacement.

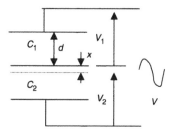

Figure 3.5. In a differential capacitor, the upper and lower plates are fixed while the center plate is movable. The capacitances of C_1 and C_2 change due to the displacement of the center plate.

3.3.4 Fiber-optic pressure sensor

Compared with other sensors, fiber-optic sensors have several advantages. They have a high sensitivity, versatility in shapes, immunity to electromagnetic interference and ability for simultaneous multiple sensing. They do not corrode inside the body and have no direct connection to external electric equipment. Thus they are safer than other sensors.

A fiber-optic system usually consists of a light source (LED, laser or infrared), one or several optical fibers, and one or several photodetectors. The measurements can be based on reflection, transmission, fluorescence, intensity change and phase change of the light. Fiber-optic methods have very high sensitivity to the picometer (10^{-12} m) level. Some of the systems are sophisticated and costly. Here we describe two simple and economic methods, which still satisfactorily measure physiological pressure.

One of the methods is based on the reflection of the light. The light comes from one fiber and is reflected from the diaphragm. The reflected light is collected by another fiber. The intensity of the collected light is dependent on the geometric configuration of the two fibers and the reflection diaphragm. When pressure displaces the diaphragm, its curvature changes the intensity of the reflected light, which is measured by photodetectors in the system. The measured light intensity has a certain relationship with the pressure (Wang *et al* 1996). Usually when an optical fiber is inserted into the body, the light intensity may change due to the bending of the fiber, temperature and other effects. A reference fiber may be bundled with the detection fiber but has total reflection at its end. It can be used as a calibration signal and reduce the error of the measurement (Tohyama *et al* 1998).

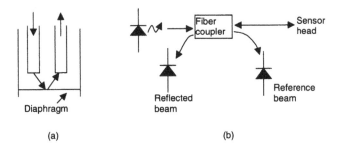

Figure 3.6. (*a*) Light comes from one fiber and is reflected by the diaphragm. The reflected beam, which goes into the outgoing fiber, is dependent on the distance between the fiber and the diaphragm and the distance between the two fibers. (*b*) Schematic diagram of the measurement system. A reference is used to avoid geometrical and other changes in the light transmission of the fibers.

Another method is based on the classic Fabry–Pérot interferometer design (figure 3.7). Typically light reflects between the optical fiber end and the diaphragm. The optical fiber end is partially transparent and is fixed. The diaphragm is totally reflective and capable of spatial translation under pressure. Interference between multiple reflections within the cavity produces *fringes* in the optical output, which is a sensitive function of the cavity's optical path length (the distance between the diaphragm and the optical end). A simple way to calculate the diaphragm deflection is to count the interference fringes. A more accurate method is to measure the intensity of the interferometer's light to calculate the distance between the diaphragm and optical fiber end. For example, when the distance is 1/4 of the wavelength, two consecutive reflective light beams are out of phase and cancel each other. Thus the output of the interferometer is 0 or minimum.

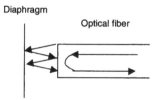

Figure 3.7. A Fabry–Pérot interferometer counts the fringes caused by diaphragm displacement.

Fiber-optic pressure sensors are widely available. A commercial product from RJC Enterprises, (Woodinville, WA) has a pressure range from –30 to 300 mmHg with an accuracy of 1 mmHg. The optical fiber has a diameter of 0.17 mm and the sensor size is 0.3 mm.

3.4 CATHETER-TYPE PRESSURE SENSORS

It is desirable to make pressure measurement noninvasively. But for many cases, a catheter must be inserted to the pressure measurement location. The sensor can be attached either at the end of the catheter (outside of the body) or at the tip of the catheter (inside the body). These two configurations have different performances.

3.4.1 Catheter-sensor pressure sensor

One type of catheter is a hollow tube and the sensor is at the external end of the catheter, which is outside the human body. The catheter is filled with fluid (usually saline) to transmit internal pressure in the body to the external sensor. The catheter is very cheap and is disposable. Older sensors were reusable, whereas modern sensors are disposable.

Due to the effect of gravity on the fluid, the sensor should be placed at the same level as the catheter tip. Otherwise, the hydrostatic head may cause error in the pressure measurement. A more practical way is to put the sensor at some reference plane, and the pressure is measured with respect to the reference level, regardless of the exact level of the catheter tip. But this still may cause some problems since the densities of the fluid in the catheter and in the human conduits are usually different. For saline in the catheter, the density is about 1.009 g ml^{-1}, while blood in the body is about 1.055 g ml^{-1}. So the contribution of gravity to the pressure is not the same and an error is inevitable.

Another problem with this kind of configuration is its dynamic response. Since most pressure measurements are static pressure measurements, the dynamic waveform is not important and this configuration is satisfactory. But in some cases, the waveform is important for diagnosis, such as for cardiovascular measurements. The waveform at the tip should be transmitted by the catheter without significant waveform distortion. Usually the velocity of the wave is fast and is not a problem. The delay through a 1 m catheter is about 2.5 ms.

Waveform distortion is often observed and is more serious. When different frequency components of a waveform propagate in the catheter, they have different amplitude attenuation and phase shift when they arrive at the diaphragm. These components sum up again at the sensor and the waveform is usually different from the original one if the system response is not flat with linear phase. The system can be modeled as a rigid tube connected to a compliant chamber filled with an incompressible viscous fluid. The input is at the catheter tip and output is volume displacement of the chamber. This kind of system usually has a natural resonant frequency and a damping ratio (Togawa *et al* 1997), which is similar to a whistle. When the input frequency is far smaller than the natural frequency, the system has a flat response. But if the input frequency is close to the natural frequency, it causes resonance and the amplitude of resonance is dependent on damping. Small damping causes a high amplitude at resonance.

The catheter sensor should have a flat frequency response over its frequency range. This requires a natural resonant frequency that is much higher than the highest frequency component in the signal, and an adequate damping of about 0.6–0.7. For cardiovascular pressures, the highest frequency component is about 50 Hz. With careful design, it is not difficult to achieve a natural frequency to satisfy this requirement. Most of the measurement systems have damping far below 0.6 (usually 0.05) and strong resonance appears. This situation can be improved by different methods, such as (1) an adequate-sized catheter, (2) mechanical damper, (3) filter or electronic compensation, or (4) numerical calculation by a computer.

3.4.2 Catheter-tip pressure sensor

A catheter-tip pressure sensor has the pressure sensor at the tip of the catheter. It is designed for accurate recording during cardiovascular pressure measurement. Since the sensor is at the tip, there is no time delay of the measurement and no error due to the reference level. It avoids the dynamic problem of the catheter-sensor configuration and the flat response increases to several kilohertz. It is minimally affected by the mechanical motion of the catheter. But a catheter-tip pressure sensor is difficult to calibrate and is relatively expensive for disposable use. With the advance of technology, either a silicon-based sensor or fiber-optic sensor may be suitable for mass production. The catheter-tip pressure sensor may become cost effective and feasible for disposable use.

The principles (strain gage, capacitance and fiber-optic) previously described can be used to design the sensor. Former studies have concentrated on the wire strain gage and bulk silicon gage. With the advance of MEMS technology, more and more new designs use the smaller semiconductor strain gage, fiber-optic element or capacitance element at the silicon tip.

The strain-gage catheter-tip pressure sensor is used in clinical applications. Figure 3.8 shows a typical design of this kind of sensor. The catheter has a vent tube to connect the rear side of the silicone rubber diaphragm to the air so the pressure reference is atmospheric pressure. The pressure sensor is attached to the back side of the diaphragm to measure the deformation of the silicone rubber diaphragm. A bundle of wires connects the sensor and other electric equipment through the vent tube. A commercial type MIKRO-TIP (Millar Instruments, Houston, TX) catheter-pressure sensor series was developed based on this design. The catheters are manufactured in different sizes from 1.3 F to 3 F (0.43–1.0 mm). The pressure range is from –50 mmHg to 300 mmHg and the sensitivity is 5 μV (V mmHg)$^{-1}$. The natural frequency is 10 kHz or higher, dependent on the size of the sensor.

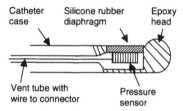

Figure 3.8. The catheter-tip pressure sensor uses a silicone rubber diaphragm for biocompatibility.

Fiber-optic catheters for pressure measurement have been studied by many investigators. The diaphragm is made using silicon as a reflector and light transmits in the fiber. These sensors are very small and the assembly is simple. The measurement system is relatively simple (section 3.3.4) and the performance is excellent. It is immune to common electromagnetic noise in clinical situations (electrosurgery, MRI, radio frequency, microwave, etc). Its low cost makes it cost effective and suitable for disposable use. A commercial catheter (Model 110/140) made by Camino Laboratories (San Diego, CA) has a pressure range from –10 to 250 mmHg. The outer diameter of the catheter is 1.3 mm.

PROBLEMS

3.1   Sketch a block diagram of an automated blood-pressure measurement system. Show all pneumatic connections, sensor locations, and briefly describe an algorithm that detects systolic and diastolic pressures.

3.2   Write the simple equation used to calculate force in applanation tonometry.

3.3   Sketch the circuit diagram for a Wheatstone bridge, as used in a silicon strain gage. For 5 V excitation and 1% change in all resistors, calculate the change in output voltage.

3.4   Modify the circuit of the capacitive pressure-sensor amplifier to provide the required bias current for the operational amplifier.

3.5   Sketch the block diagram of the fiber-optic pressure sensor showing all fibers and all detectors.

3.6   Sketch the amplitude and phase of the frequency response of a catheter pressure sensor with the pressure sensor outside the body.

3.7   A catheter-tip pressure sensor has 5 V excitation. Calculate the required amplifier gain to increase the sensor output for a 0–300 mmHg input range to the 0–5 V required by an analog-to-digital converter.

REFERENCES

Pallás-Areny R and Webster J G 2001 *Sensors and Signal Conditioning* 2nd edn (New York: John Wiley)

Peura R A 1998 Blood pressure and sound *Medical Instrumentation: Application and Design* 3rd edn ed J G Webster (New York: John Wiley)

Thakor N V, Webster J G and Tompkins W J 1984 Estimation of QRS complex power spectra for design of a QRS filter *IEEE Trans. Biomed. Eng.* **31** 702–5

Togawa T, Tamura T and Oberg P 1997 *Biomedical Transducers and Instruments* (Boca Raton, FL: CRC)

Tohyama O, Kohashi M, Sugihara M and Itoh H 1998 A fiber-optic pressure microsensor for biomedical applications *Sensors Actuators A* **66** 150–54

Wang A, Miller M S, Plante A J, Gunther M F and Murphy K A 1996 Split-spectrum intensity-based optical fiber sensors for measurement of microdisplacement, strain, and pressure *Appl. Opt.* **35** 2595–601

Wu N, Hu M, Shen J and Ma Q 1993 A miniature piezoresistive catheter pressure sensor *Sensors Actuators A* **35** 197–201

CHAPTER 4

X-RAY-BASED IMAGING

Dean H Skuldt

In 1895, a discovery that allows the study of a patient's internal anatomy using X rays forever changed the nature of medicine. This revolution, begun by Roentgen, enabled physicians to perform diagnoses without ever physically, surgically invading patients.

As we move into a new realm of minimally invasive medicine, we see our ability to image anatomy and function at the center of this progression. This section presents brief physical descriptions of each modality, and the new advances in the direction of noninvasiveness of five separate modalities: optical imaging, X-ray-based imaging, nuclear medicine, magnetic resonance imaging (MRI) and ultrasound.

Even after more than a century has passed since the initial discovery of X rays by Roentgen, X rays are still at the heart of most of the imaging used in medicine today. In all X-ray imaging, a beam of X-ray photons passes through the anatomical region of interest, and produces an image from the transmitted photons. Technological developments have greatly reduced the radiation exposure to patients in many procedures, and the introduction of digital methods for X-ray-based imaging provide clinicians with views that were never before available, and change the way in which X-ray-based images are acquired, processed and stored. Some procedures require a higher level of radiation to the patient, but the advantages in superior image quality and/or utility usually outweigh the higher risk of radiation. This chapter discusses X-ray production, the interaction of X rays with matter, X-ray detection, mammography, fluoroscopy, computed tomography (CT) and CT angiography.

4.1 X-RAY PRODUCTION

X-ray photons are created by accelerating electrons to a high velocity within an X-ray tube, and allowing them to collide with their metal target. This section briefly describes the spectrum of the X-ray beam, the X-ray tube and anode design.

4.1.1 The X-ray beam

Electrons that hit a tungsten target material at high velocity produce two types of photon radiation: brehmsstrahlung and characteristic radiation. Brehmsstrahlung is produced when an electron decelerates by travelling near the nucleus of a target atom. In figure 4.1, photons cause the higher energy portion of the curve in the plot of photon energy. Characteristic radiation is produced when an electron causes an electron of a particular atomic shell in the target to be displaced. The displaced electron is replaced by a target electron of higher energy from an outer shell. This transition from shell to shell produces radiation of a particular energy, and appears as a spike on the plot of photon energy.

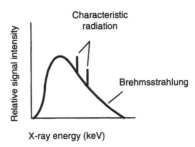

Figure 4.1. X-ray photon energy consists of both brehmsstrahlung, shown by the upper portion of the smooth curve, and characteristic radiation, which appears as spikes on the plot.

4.1.2 X-ray tubes

An X-ray tube consists of an anode (target), a cathode, and is contained inside a vacuum (figure 4.2). The dc voltage between the anode and the cathode is typically in the kilovolt range. An X-ray system therefore requires a high-voltage generator. The cathode is a coil filament (typically made of tungsten) that is heated by forcing a current through the coil. When the cathode has reached a high temperature (typically around 2000 °C), the electrons on the cathode begin to boil off. When the high voltage between the anode and cathode is applied, the electrons are drawn to the anode. The tube current (mA) is determined by the rate at which the electrons boil off of the cathode, and it is therefore a function of the filament temperature. The velocity and energy of the electrons before they hit the anode are functions of the tube potential (kilovolts peak—kVp). Tube current determines the amount of exposure (along with duration of exposure), and tube potential determines the hardness, or penetrating ability of the radiation.

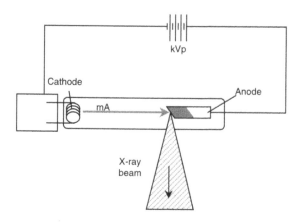

Figure 4.2. An electric current heats the cathode and boils electrons from it. These are drawn to the anode by a high voltage between the cathode and anode. Electrons that impinge the target produce an X-ray beam, which is directed out of the tube.

4.1.3 Anode design

Depending on the application, X-ray production requirements vary in intensity, duration and level of resolution. In imaging moving structures, a very short duration of high intensity is required, while imaging of very small structures requires high resolution. These imaging requirements are limited by anode heat dissipation.

Resolution is a function of beam output and focal spot size (the area of the anode that is impinged by electrons). Unfortunately, a smaller focal spot increases heat production, high levels of which may damage the anode. In addition, heat production varies directly with tube current and duration. The efficiency of X-ray production and hardness of radiation is a function of kilovolts peak (kVp). Acceptable levels of all of these parameters must be found for a given application.

Only a small part of the energy transmitted from the anode is useful radiation. Ninety-nine percent of the energy that results from the electrons impinging the target is transformed into heat. Most modern X-ray tubes use rotating anodes, which help to minimize the heat dissipation in a single area of the anode.

4.2 INTERACTION OF X RAYS WITH MATTER

Only 1–4% of X rays travelling into the body reach the detector as primary radiation. The rest of the X rays interact with tissues by either being scattered by Compton or photoelectric effects or absorbed. This section discusses these effects and their impact on the health of patients.

4.2.1 Scattering

Two types of interaction between X rays and biological tissue cause X rays not to be detected, serving as the basis for a diagnostic image. The Compton effect causes an electron to be scattered, so it never reaches the detector plane. The photoelectric effect results in complete absorption of energy.

The relative probability of the two effects taking place depends on the magnitude of the X-ray energy used. Therefore, the type of energy used is chosen to provide the best image for the certain tissue. The number of Compton events is not related to the atomic number of matter, but it increases as mass density increases (bone vs soft tissue). As energy increases, the probabilities of Compton events are nearly the same. As energy decreases, their relative probabilities become different.

The probability of Compton-scattering events is inversely proportional to the X-ray energy, whereas the probability of photoelectric-effect events is inversely proportional to the cube of X-ray energy. Therefore, as X-ray energy increases, Compton events dominate.

4.2.2 Harmful effects of exposure

As X-ray photons are sent through the body, they are either scattered or absorbed by tissue. Whether this happens by photoelectric or Compton events, atomic electrons are liberated and subsequently sent through the tissues as well. These liberated electrons may go on to interact with hundreds or thousands of other molecules, including DNA molecules. Other molecules (including water) may be turned into free radicals by the electrons, which in turn, may interact with DNA. These interactions with DNA often lead to damage that is either insignificant or reparable. But in other cases, damage may lead to carcinogenesis, genetic defects in offspring, or congenital abnormalities in the case of irradiation to a fetus. For this reason, exposure to ionizing radiation should be kept to a minimum.

The danger of ionizing radiation is a concern not only for patients, but also for medical staff. In procedures that require staff to be exposed regularly, such as interventional procedures, staff may meet regulatory limits of exposure, requiring them to no longer work for a given period of time. These factors drive the need to eventually replace many X-ray-based applications with less- or noninvasive imaging alternatives.

4.3 X-RAY DETECTION

Until recently, X-ray images were produced by exposing either a phosphor screen in the case of fluoroscopy, or a screen and film. Today images may be acquired digitally using a number of different detector technologies, some of them digital, allowing the benefits of digital image processing. Figure 4.3 shows a schematic of an X-ray-based examination. In this case, a screen–film combination detector is shown, but the figure may be generalized with the screen–film combination detector replaced with a number of different detector types.

The method of detecting transmitted X rays in a particular application depends on the specific requirements of the application. Common detection methods, screen–film combination, image intensifier and digital detectors, are briefly described following a discussion of image quality.

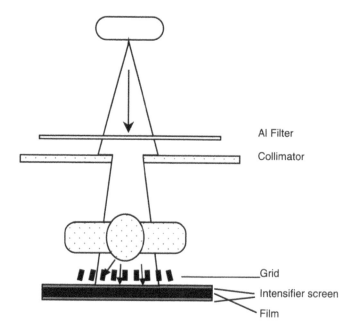

Figure 4.3. X-ray photons emitted from an X-ray tube first pass through an aluminum filter to reduce the low-energy radiation, then through a collimator, which limits the beam to a small area, and through the patient. A collimator grid reduces scattered radiation, and a detection medium absorbs the transmitted radiation. In this case a screen–film combination is used. Intensifier screens on both sides of the film increase the intensity of the light before it reaches the film.

4.3.1 Screen–film detectors

Today the most widely used X-ray detector is the screen–film combination detector. Image intensifier screens placed on both sides of the film increase the signal recorded by the film. Using intensifier screens allows reduced dosage, decreasing patient exposure, but they also cause a smearing of detected photons. For this reason, in mammography only one screen is used to improve the clarity of small details.

The characteristic curve for screen–film detectors is S shaped, as shown in figure 4.4. Exposure is kept to the central portion of the curve for a linear dose-intensity response. Image contrast is determined by the slope of the linear region.

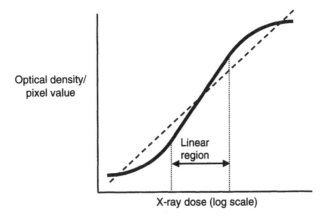

Figure 4.4. The characteristic curve for screen–film combination detectors is S shaped (solid curve), providing a relatively small dynamic range, as indicated by the linear region. Digital detectors have a linear characteristic curve (dashed line), allowing for a larger dynamic range.

4.3.2 Image intensifier

The image intensifier (figure 4.5) is the detector used in fluoroscopy, where an X-ray beam is continually applied to visualize movement in a real-time manner. X rays that interact with the input phosphor (usually CsI) release visible light photons. CsI crystals are grown as tiny pins, which, when arranged in the phosphor, provide very high resolution. Released photons from the input phosphor interact with the photocathode by a process called photoemission, which results in the release of electrons. These electrons are pulled toward the anode by 20 kV at the other end of the tube, and guided by electrostatic lenses to be focused on the output phosphor. Visible light photons released by the output phosphor may be recorded by a video camera or charge-coupled device (CCD) camera.

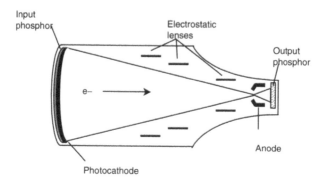

Input
phosphor

Electrostatic
lenses

Output
phosphor

e–

Anode

Photocathode

Figure 4.5. X rays that hit the input phosphor of an image-intensifier tube release visible light. Light photons react with the photocathode to produce electrons, which are guided to the anode by electrostatic lenses. Electrons that hit the output phosphor, in turn, emit visible light, which may be recorded by a video camera.

4.3.3 Digital detectors

In many applications, digital detectors may replace other detection methods. Despite relatively low spatial resolution compared to other methods, digital detectors offer many advantages.

In contrast to screen–film combination detectors, a digital detector provides a large dynamic range as a result of its linear characteristic curve (figure 4.4). Digital detection allows the use of digital image-processing algorithms for contrast enhancement, combination of images (digital subtraction angiography), spatial-frequency processing, dynamic range reduction, and multiscale contrast amplification. The display of digital images may be modified while being viewed, providing the viewer with different ways of viewing different details. Because digital detection has a large dynamic range, fewer retakes are caused by over- and underexposure, reducing the level of patient exposure.

The quality of a digital image is determined by the pixel size, pixel depth, dynamic range and signal-to-noise ratio (SNR). By the Shannon theorem, if the pixel size is much smaller than the smallest image detail, there is no loss of information. In applications that require resolving very small detail, such as mammography, digital detection currently cannot meet the performance of film-based detection.

Digital detectors come in several different forms. Image intensifiers with CCD cameras, storage phosphor screens, and solid-state selenium drum detectors are either being used clinically or being developed. A recently developed type of digital detector is the flat panel detector, which consists of an X-ray sensitive layer and an active detector matrix (Kamm 1997). Flat panel detectors are of either direct or indirect type. Using the direct method, X rays are converted to charge by a semiconductor layer of selenium, and the active detector detects electric charge. Using the indirect method, X-rays are converted to light by a scintillator layer, and the detector detects the light and converts it into electric charge. In either case, activated cells in the two-dimensional matrix correspond to a given pixel location.

4.4 IMAGE QUALITY

Generally, the quality of an imaging system is described by the modulation transfer function (MTF). This section briefly describes the MTF, and the factors that contribute to X-ray image quality: spatial resolution, contrast and sensitivity.

The MTF indicates the image-to-object contrast as a function of spatial frequency. Ideally, we desire a system that gives 100% contrast at all spatial frequencies, but a real imaging system loses contrast as spatial frequency is increased (figure 4.6).

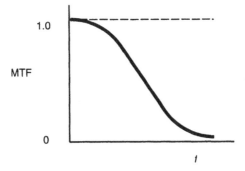

Figure 4.6. An ideal modulation transfer function would be one which transfers 100% contrast at all frequencies (dashed line). In reality, an imaging system transfers contrast at low spatial frequencies very well, and contrast diminishes as spatial frequency increases (solid curve).

The system's total MTF is the product of the MTF of each component in the system. For instance, in an X-ray system the MTF of the focal spot, the screen and the film are measured separately, and their product is the system MTF:

$$MTF_{system} = MTF_{FS} \times MTF_{S} \times MTF_{F}.$$

Therefore, the system MTF is always worse than its weakest component.

Spatial resolution depends on aspects of the tube design, such as the focal spot size. The smaller the focal spot, the higher the resolution. The type and quality of detector used influences the spatial resolution as well. For instance, the size of grains in films and the size of pixels in digital detectors determine the ability to resolve small detail. Typical pixel sizes for digital detectors range from

100 μm to 200 μm for phosphor plates, 120 to 200 μm for flat panel detectors, and 150 to 500 μm for image-intensifier video systems.

Contrast in images depends on factors such as the X-ray source parameters (mA and kVp), the characteristic curve of the detector, and the use of contrast agents. Depending on the type of tissue contrast desired, the hardness of radiation must be adjusted. For instance, for better soft-tissue contrast, radiation must be less penetrating. As discussed earlier, a steeper characteristic curve increases contrast.

Sensitivity, or speed, of detection is an important factor in imaging of moving objects, as it has a direct bearing on temporal resolution. Also, higher sensitivity may facilitate keeping exposure to a minimum. Sensitivity has a reciprocal relationship with resolution, meaning that as sensitivity is increased, resolution is decreased.

4.5 X-RAY APPLICATIONS

Depending on the need in a particular application, the method, delivery, detection and the type of radiation are uniquely designed. Several applications are described: mammography, fluoroscopy and X-ray angiography.

4.5.1 X-ray mammography

Detection of calcifications in the breast is an important procedure in the early detection of breast cancer. Early detection is of vital importance, as breast cancer is the leading cause of death in women by cancer, and treatment of cancer is much more effective if detected in its early stages. Today, despite continuing developments in mammography using a number of other less-invasive modalities, X-ray mammography dominates the area of breast cancer detection.

In mammography, as compared to conventional radiography in which contrast between bone, soft tissue and lung is desired, tissue contrast between soft tissues is required. Lower peak kilovoltage is used, typically 20–35 kVp, in order to maximize soft- tissue contrast. This reduces the X-ray beam, so in order to compensate, the filament current is increased, which increases patient exposure.

High resolution is required for detecting small calcifications in breast tissue. For this reason, tubes with very small focal spots are used. Also, breast compression is used to separate overlying tissues that might be superimposed in an image and decrease the tissue thickness to reduce scatter.

Digital radiography is being used for mammography with some success. Digital radiography suffers in spatial resolution compared to screen–film detectors. However, it has the advantage of increased dynamic range and the ability to use postprocessing algorithms (i.e. edge detection) and to allow computer-assisted diagnosis (Sabel and Aichinger 1996).

X-ray screen–film technology is the gold standard in mammography today due to its superior detection capability. However, X-ray mammography still has room for improvement in its sensitivity and specificity, and radiation dosage to the patient is of great concern.

4.5.2 Fluoroscopy

In order to study motion and dynamics, and to aid in patient positioning for high-intensity images, constant real-time imaging may be accomplished with a low-exposure-rate X-ray source using fluoroscopy. X-ray fluoroscopy is currently an important method for guiding minimally invasive procedures.

In a typical fluoroscopic examination room, the X-ray source is beneath the patient couch, and transmitted radiation is detected above the patient by an image-intensifier tube (section 4.3.2) and video camera, and the video images are displayed on a video monitor. Generally, a contrast agent is injected into the patient to highlight anatomy. Digital images may be acquired from fluoroscopy studies in what is known as digital fluoroscopy. This provides the advantages of digital image processing, which allows for flexibility in display, image enhancement, etc.

4.5.3 X-ray angiography

Imaging of vessels is important both for diagnosis of vascular disease, as well as for assisting in interventional procedures such as stent placement, balloon angioplasty and thrombolysis (declotting). Fluoroscopy is commonly used for catheter guidance and interventional procedures. Digital subtraction angiography (DSA), developed in the 1970s, continues to be the most widely used method for obtaining vascular images. The principles of DSA are also being used in MR angiography.

In DSA, an image acquired before the injection of a contrast agent is subtracted from images taken after injection. In this way, the resulting images highlight the vessels containing the contrast agent (Belli 1997).

4.6 COMPUTED TOMOGRAPHY

X-ray imaging from a single projection is limited in its ability to produce accurate and detailed images of anatomy, especially when the region of interest has many overlapping structures. Computed tomography (CT) uses information from multiple projection images taken at different angles, and computes an image of the region of interest from the projections.

4.6.1 Scanner technology

CT scanners have evolved through several different generations. Figure 4.7 shows two more recent types. Fourth-generation scanners, which have a moving X-ray tube and a fixed-position detector ring, are commonly used. The fifth-generation scanner, few of which exist in 2000 due to its relatively recent arrival, has no moving parts, which eliminates many of the problems with the scanners that preceded it. The beam of electrons from an electron gun is deflected by a set of magnetic coils. The electrons hit a fixed target ring, which sends X rays through the patient and to the fixed detector ring. This configuration allows very fast imaging times, and reduced patient dose.

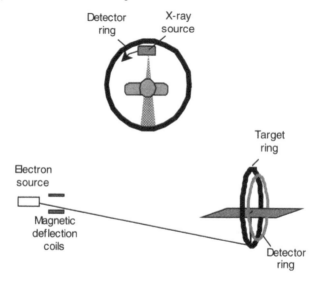

Figure 4.7. Fourth-generation scanners (top) have a rotating X-ray source and a stationary ring of detectors. Fifth-generation CT scanners (bottom) have no moving parts. An electron source is deflected by magnetic coils in order to produce X rays from a given point on the target ring. Transmitted X rays are detected by the detector ring on the other side of the patient.

4.6.2 Filtered back-projection

Early methods of reconstruction of CT were iterative methods. With this type of method, an initial guess at the image is made, and projections are used successively to correct a running result. With a large number of pixels and projections, this method took a very long time to perform. Today analytic methods, such as filtered back-projection are used.

Figure 4.8 shows the concept of back-projection. In this case, only two lines of CT data, taken at 90° from one another, are used to produce a back-projection image. A real image would be produced by back-projecting data from acquisitions at all angles.

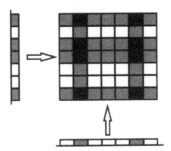

Figure 4.8. Two lines of projection data are back-projected to produce an image.

In order to produce the actual image, the image must be transformed into the spatial frequency domain, filtered by a filter proportional to the spatial frequency, and transformed back into image space. In order to save processing time, the order of these steps may be changed slightly. The individual lines of projection data may first be transformed into the 1D spatial frequency domain, then filtered. This processing can take place after the first acquisition from the CT scanner is complete. After all of the filtered lines are acquired, the entire image may be formed by back-projecting and transforming to image space. This order allows the reconstruction process to take place during the scan, providing an image much sooner after the scan.

4.6.3 Spiral CT

A relatively recent development in CT technology is spiral, or helical CT. Figure 4.9 shows that the patient is scanned in a helical configuration by allowing the patient to move through the gantry while the source and detector rotate around the patient continuously.

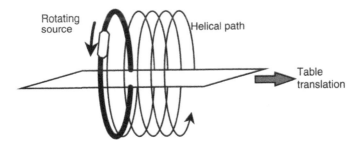

Figure 4.9. In spiral CT, helical scan trajectory is accomplished by longitudinal translation of the patient and continuous rotation of the source.

Spiral CT allows fast imaging times, and provides improved images of anatomy, which would otherwise be degraded by respiratory motion. This is the case particularly in chest, abdomen and pelvic scans.

PROBLEMS

4.1   Distinguish between characteristic radiation and bremsstrahlung.
4.2   Explain how the X-ray beam intensity is controlled.
4.3   Describe two effects that degrade X-ray image quality.
4.4   Explain the reason for using the aluminum filter, collimator, grid and intensifier screens.
4.5   Compared to screen–film detectors, how does the characteristic curve of a digital detector decrease patient exposure?
4.6   Describe all conversions in an image intensifier from X ray input to light output.
4.7   Describe the difference between a fourth- and fifth-generation CT scanner.
4.8   Back-project the following two lines of data, acquired orthogonally to one another, to give an indication of the signal level of the contents of the original two-dimensional grid: [1 4 8 3 2];   [9 5 5 1 1].
4.9   Explain how CT image reconstruction is accomplished.

REFERENCES

Ammann E and Kutschera W 1997 X-ray tubes—continuous innovative technology *Br. J. Radiol.* **70** S1–9
Belli A-M 1997 The future of arteriography and vascular interventional radiology *Br. J. Radiol.* **70** S168–70
Bushong S C 1997 *Radiologic Science for Technologists: Physics, Biology and Protection* 6th edn (St Louis, MO: Mosby)
Dixon A K 1997 The appropriate use of computed tomography *Br. J. Radiol.* **70** S98–105
Hansell D M 1997 Thoracic imaging—then and now *Br. J. Radiol.* **70** S153–61
Kamm K F 1997 The future of digital imaging *Br. J. Radiol.* **70** S145–52
Rees M 1997 Cardiac imaging: present status and future trends *Br. J. Radiol.* **70** S162–7
Richenberg J L and Hansell D M 1997 Image processing and spiral CT of the thorax *Br. J. Radiol.* **70** S708–16
Sabel M and Aichinger H 1996 Recent developments in breast imaging *Phys. Med. Biol.* **41** 315–68
Wolbarst A B 1993 *Physics of Radiology* (Norwalk, CT: Appleton & Lange)

CHAPTER 5

NUCLEAR MEDICINE

Dean H Skuldt

Nuclear medicine is a method of using radionuclides, radioactive pharmaceutical agents that bind to specific sites or organs in the body, in order to image organ functions. This modality involves injecting a radionuclide into the patient, allowing time for it to be taken up by the organ of interest, and then detecting the spatially and temporally dependent signal given off by the radionuclide bound to the organ. This chapter discusses radionuclides, gamma-ray detection, and aspects of the two major types of imaging in nuclear medicine, positron emission tomography (PET) and single-photon emission computed tomography (SPECT).

5.1 RADIONUCLIDES

Nuclides consist of any combination of protons and neutrons that comprise an atom. About 280 of the 1500 nuclides are stable, and the remainder, which are unstable, or radioactive, are called radionuclides.

Stable atoms have a neutron-to-proton ratio that is close to one for light elements, and close to 1.5 for heavy elements. A radionuclide has a neutron-to-proton ratio that is either above or below this *line of stability*. In order to become more stable, a radionuclide undergoes a process of decay, whereby subatomic particles are emitted. Every radionuclide decays in a unique way, involving a characteristic amount of energy.

In nuclear medicine, radionuclides that are of use are those that undergo either positron decay or gamma decay. Gamma photons may be detected in both SPECT and PET imaging, but different radionuclides are used for each. SPECT uses radionuclides that emit gamma photons. PET uses radionuclides that emit positrons, which, in turn, annihilate with electrons to produce two gamma photons.

An important characteristic of radionuclides is the rate at which they decay. The equation governing the rate of decay is

$$A = A_0 e^{-\lambda t} \tag{5.1}$$

where A is the activity at time t after a starting time, A_0 is the activity at the starting time and λ is the decay constant. Decay rates are indicated by their half-life, $t_{1/2}$, which is related to the decay constant by

$$\lambda = \frac{0.693}{t_{1/2}} . \tag{5.2}$$

The SI unit quantifying the level of radioactivity is the bequerel (Bq). One Bq is the amount of radiation in a sample that decays at a rate of 1 dps (disintegration per second).

Radionuclides made from particular elements may be incorporated into particular molecules. These molecules, which are injected into or inhaled by the patient, are those that bind to particular tissues in the body. It is the amount of radiation emitted by these molecules that is the detected quantity in nuclear medicine. The differential uptake or usage of these molecules by particular tissues or organs gives an indication of function.

As with X rays, gamma rays emitted by radionuclides are ionizing, and their effects on human tissue are damaging. Therefore, like X-ray applications, dosage in nuclear medicine must be kept to a minimum but must be high enough to produce useful images in a scan duration of reasonable length.

5.2 GAMMA DETECTION

The earliest devices used in nuclear medicine did not produce images. A detector was used simply to count the emission of gamma rays from the surface of the body, with no intention of resolving the spatial distribution of radionuclide concentration. The first imaging devices used a single detector, which scanned a rectilinear path in order to produce a planar image from the line-by-line data from the detector. A further advancement of nuclear medicine technology was brought about with the introduction of the gamma camera.

Figure 5.1 shows that a gamma camera consists of a multihole lead collimator, a crystal scintillator, typically made of NaI, and a photomultiplier tube (PMT) array. Gamma radiation from the organ of interest that hits the crystal causes the emission of visible light, which is detected by the PMT. A multihole collimator allows only those gamma rays that are incident along parallel paths to be detected. The transverse (*x* and *y*) location of an emitting radionuclide is determined by the particular PMT that detects the event, and the depth is determined by analyzing the amplitude of the detected pulse. Spatial resolution is typically 1 cm in a gamma camera.

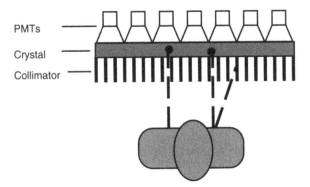

PMTs

Crystal

Collimator

Figure 5.1. In a gamma camera, a scintillation event in the crystal is detected by a photomultiplier array. A multihole collimator restricts detection to those photons that travel directly through the parallel collimator leaves.

5.3 SINGLE-PHOTON EMISSION COMPUTED TOMOGRAPHY

A limitation in using a single gamma camera is that objects that overlap one another and overlying tissue cause attenuation, and subsequently, reconstruction problems. A SPECT scanner overcomes this limitation by using between one and four gamma cameras that rotate about the patient slowly to collect gamma radiation. In this way, projections of detected photons from a multitude of different angles may be used to reconstruct images using filtered back-projection (section 4.5.2). Figure 5.2 shows the advantage of SPECT over planar images.

Using SPECT, quantification of radionuclide concentration is typically not possible due to poor sensitivity. Instead, images provide a qualitative image, showing relative concentrations of the radionuclide distribution. However, some currently available scanners provide attenuation correction, which allows quantitative assessment. In addition, despite poor computation time, iterative reconstruction methods appear to give more accurate results (Medley and Vivian 1997).

SPECT is used in neuroimaging to assess cerebral blood flow (rCBF) and changes in neurotransmitter receptors (Müller-Gärtner 1998), and in cardiology to assess glucose metabolism. SPECT is commonly used to screen for bypass surgery using thallium-201, where image intensity is indicative of cardiac perfusion (Ollinger and Fessler 1997).

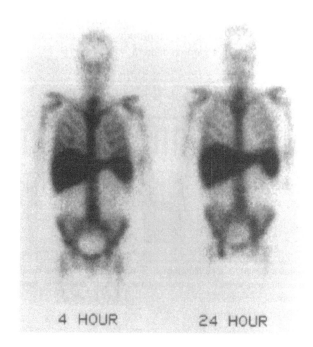

Figure 5.2. The planar images (top row) are obtained early (4 h prior injection) and late (24 h prior injection) and demonstrate the expected biodistribution of the patients own white blood cells (WBC) labeled with indium-111 and reinjected. The liver, spleen and bone marrow regions are visualized, and in the 24 h image the right groin region and deep within the pelvis a left-sided abnormality are just discernible. The lower images represent 'stacked' transverse slices of SPECT oriented directly at the reader (left lower) and about 60° to the left of the reader (right lower). These images demonstrate the expected pelvic bone marrow activity and linear uptake of the right femoral and iliac arteries and the left iliac artery. This indicates the presence of In-111-WBC, and so confirms the vascular graft that this patient has is infected. Whereas the planar images suggest infection in parts of the graft, the SPECT images confirm extensive disease. Courtesy of Michael A Wilson, MD, University of Wisconsin-Madison.

Recent advances in SPECT include the advent of ^{99m}Tc-labeled agents for cardiac perfusion, which are much more sensitive than thallium, its predecessor, and the introduction of 511 keV gamma cameras, which are capable of detecting emissions from positron-emitting radionuclides (Rees 1997).

5.4 POSITRON EMISSION TOMOGRAPHY

In positron emission tomography, radionuclides that cause the emission of positrons are localized by detecting annihilation events that emit two photons in opposite directions. This method, like SPECT, provides functional information based on the uptake of radionuclides by specific tissues. Compared to SPECT, PET has the advantages of increased signal intensity, faster scan times, the ability to produce quantitative maps of radionuclide distribution, and the elements that may be made to emit positrons occur naturally, and therefore are more easily incorporated into molecules that may be used as tracers. However, PET scanners are very expensive and require on-site (or quickly accessible) radionuclide production due to short half-lives. Nonetheless, PET finds particular utility in diagnosing cancer and studying organ function. This section discusses event detection, image reconstruction and uses of PET.

5.4.1 Event detection

When a positron is emitted within the body, it travels typically less than a millimeter while it interacts with other particles, and finally annihilates by colliding with an electron. This collision causes two 511 keV photons to be emitted, both of which travel in exactly opposite directions.

In a modern PET scanner, a ring of detectors surrounds the patient. The position of an annihilation event is detected, not by using collimators, but by means of coincidence timing. Figure 5.3 shows that when an event is detected in two detectors within a very small timing window, called the coincidence timing window, this event is considered to have occurred along the straight line between the two detectors, and is called a coincidence detection. The line connecting the two detectors in coincidence detection is called a line of response (LOR).

When an event is scattered and both photons are detected, their detection may occur within the coincidence timing window, but the annihilation does not occur within the LOR. In order to reject a scattered event, the intensity recorded at the detectors is analyzed. When photons scatter, they lose part of their energy. Therefore, events are rejected if their intensity is below a particular threshold.

Another type of event, in which only one photon is detected and the other photon is not recorded by any detector, is called a single. An accidental occurrence may result if two singles are recorded within the coincidence timing window. The rate of occurrence of these events may be reduced by shortening the coincidence timing window, and it is increased by radionuclide concentration. Accidental coincidences are sometimes accounted for by statistical means during image reconstruction.

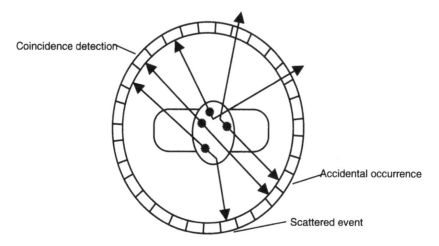

Figure 5.3. In a PET scanner, three types of photon detection resulting from annihilation (dark dots in the figure) may occur within a single axial plane. A coincidence detection is the result of an annihilation event along the LOR between the two detectors. A scattered event is detected by two detectors whose LOR does not include the position of the annihilation that was scattered. An accidental occurrence is the result of two singles (events in which only one photon is detected and the other misses the detector ring in the axial direction) that occur within the coincidence timing window.

This method of detection in PET, since no collimators are used, provides a higher level of detection than SPECT. This higher level of detection, which increases image statistics, allows PET to produce quantitative images of radionuclide concentration.

From a given area in the photon-emitting organ, lines of response are produced between many detectors. As in CT, these lines of response are used to reconstruct an image using filtered back-projection.

5.4.2 Uses of PET

Radionuclides that emit positrons may be produced from oxygen, nitrogen, carbon and other elements. These may be prepared as gaseous molecules of O_2, CO, CO_2 or NH_3, which may be incorporated into other molecules that are important in particular biochemical pathways. O_2-containing molecules are used for monitoring blood flow and oxygen metabolism.

In brain imaging with PET, a popular molecule used is a glucose molecule labeled with F^{14}. Since this molecule crosses the blood–brain barrier and stops at the position where it is first taken up by tissue, this molecule is used to map glucose uptake in the brain.

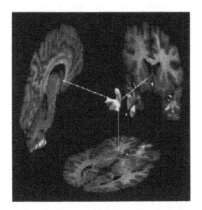 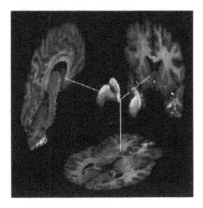

Figure 5.4. Spatial correlation (co-registration) of anatomic and physiologic (functional) images of the brain. Each of the three planar images (left and right) is derived from a high-resolution 3D scan of the brain using MRI (3D spoiled gradient echo sequence). These provide the anatomic framework of the whole brain, showing the location of the structures being measured by the physiologic images. The surface-rendered volumes at the intersection of the crosshairs for both the normal volunteer (right) and the Parkinsonian patient (left) are biochemical images of neurotransmitter function obtained by PET scanning using the labeled precursor 6-[18F]fluoro-3,4-dihydroxyphenylalanine (L-DOPA). The nerve cells affected by Parkinson's disease and related disorders normally produce the neurotransmitter dopamine from natural L-DOPA, and store it in nerve terminals in the central part of the cerebrum. The volume of the surface-rendered objects is proportional to the number of these dopaminergic cells that are active or healthy. The surface-rendered structures in the right-hand image show the shape of normal dopamine accumulation. In Parkinsonian patients many of these specific nerve cells have died out. Since these dopaminergic nerve cells have very little volume, their loss does not significantly affect the MRI appearance; but PET imaging with radio-labeled L-DOPA shows a drastic reduction in the accumulation of labeled dopamine (left). Courtesy of Doug Brown, MD and Terry Oakes, PhD, University of Wisconsin-Madison.

PET imaging is not a widely used imaging modality due to the high cost of scanners, and the need for a local cyclotron (for producing radionuclides with short half-lives), which also comes at a very high cost. However, it provides quantitative images of function that are not attainable by any other modality.

5.5 IMAGE QUALITY

In both SPECT and PET, images are reconstructed using filtered back-projection (section 4.5.2), and the quality of images depends on some of the same parameters. For both modalities, image quality is determined, in part, by the following: spatial resolution of the scintillation camera, spatial linearity (the ability to produce a straight image from a straight object), energy resolution (precision of the photon-energy measurement), flood-field uniformity (the ability to produce a uniform image from a uniform distribution of photons), sensitivity (number of photons recorded per unit source radioactivity), and count-rate linearity (the ability of the camera to record a count rate proportional to the event rate).

PROBLEMS

5.1   A sample contains 42 MBq of ^{201}Tl, which has a half-life of 73 h. Calculate the radioactivity after 16 h.

5.2   Describe the three types of detected event in PET and how they can be distinguished from one another.

5.3   We have a PET system that has a ring of 180 detectors evenly spaced about a ring whose radius is 1 m. The detectors are numbered 1 through to 180 increasing clockwise, with detector 1 at 2°, detector 2 at 4°, and detector 180 at 0°. The coincidence timing window is 1 ns. The following photon detections were made:

> detector 15   @ $t = 4.0$ ns;
> detector 100 @ $t = 5.5$ ns;
> detector 160 @ $t = 6.0$ ns.

Find the equation for the line along which the positron-emitting source lies.

More detections are made:

> detector 3   @ $t = 18.0$ ns;
> detector 121 @ $t = 18.1$ ns.

What are the coordinates of the positron-emitting source?

5.4   Describe the types of clinical measurement made with SPECT imaging.

REFERENCES

Bernier D R, Christian P E and Langan J K 1994 *Nuclear Medicine Technology and Techniques* (St Louis, MO: Mosby)

Dobbins J T III, Hames S M, Hasegawa B H, DeGrado T R, Zagzebski J A and Frayne R 1999 Medical imaging *The Measurement, Instrumentation and Sensors Handbook* ed J G Webster (Boca Raton, FL: CRC)

Links J 1998 Advances in nuclear medicine instrumentation: considerations in the design and selection of an imaging system *Eur. J. Nucl. Med.* **25** 1453–66

Medley C and Vivian G 1997 Radionuclide developments *Br. J. Radiol.* **70** S133–44

Müller-Gärtner H 1998 Imaging techniques in the analysis of brain function and behaviour *Trends Biotechnol.* **16** 122–30

Ollinger J J and Fessler J A 1997 Positron-emission tomography *IEEE Signal Process.* **14** (1) 43–55

Rees M 1997 Cardiac imaging: present status and future trends *Br. J. Radiol.* **70** S162–7

Wolbarst A B 1993 *Physics of Radiology* (Norwalk, CT: Appleton & Lange)

CHAPTER 6

MRI

Dean H Skuldt

As we compare the invasiveness and imaging power of medical imaging modalities, it is magnetic resonance imaging (MRI) which appears to win the proverbial prize. Apart from its relative noninvasiveness compared to other modalities that use ionizing radiation, its ability to provide soft-tissue differentiation is superior to any modality, and images may be acquired in any direction, as it does not rely on transmission and attenuation through the subject. In addition, MRI offers tremendous flexibility in its ability to not only image anatomy, but also to image functions.

This chapter describes the physical basis of the MR signal, the techniques used to create an MR image, and the factors which contribute to contrast and resolution. New developments in the field include MR angiography, diffusion-weighted and functional MRI, and MR spectroscopy.

6.1 MR PHYSICS

An understanding of an imaging modality based on nuclear magnetic resonance requires an introduction to the physics in which it is rooted. This section describes briefly the concepts of precession, excitation and relaxation.

6.1.1 Precession

Atoms that have an odd number of protons or neutrons (odd spin quantum number, I) have the property of spin angular momentum. These atoms are participants in the MR phenomenon, and they are referred to as *spins* as they may be envisioned as spheres of spinning charges creating a current density. This current surrounding the nucleus results in a magnetic moment, $\mu = \gamma J$, where γ is the gyromagnetic ratio for the particle, and J is the angular momentum of the particle.

The most common atomic species used in MR imaging of biological tissue is ^{1}H (a single proton) because of its high relative abundance and high sensitivity (it creates a stronger MR signal). ^{1}H has a gyromagnetic ratio λ of 267.53 MHz T^{-1}. This ratio, as shown by the Larmor relation equation (6.1), relates the Larmor or precessional frequency f_0 to the external magnetic field strength, B.

$$f_0 = \frac{\gamma}{2\pi} B . \tag{6.1}$$

Precession is analogous to the way a top, while spinning, precesses about the vertical axis. Figure 6.1 shows that as protons precess in the presence of an external magnetic field, they may do so in two possible energy states: either excited (oriented antiparallel to the external magnetic field) or relaxed (aligned with the magnetic field).

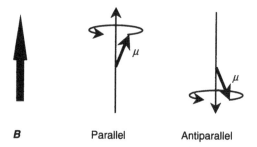

Figure 6.1. In the presence of an external magnetic field B, a precessing spin μ may be oriented either parallel or antiparallel to the direction of the external field.

As we consider MR imaging, we may focus the discussion on a single voxel (volume picture element), a three-dimensional volume containing a population of spins. These spins together exhibit a net magnetic moment of zero when no external magnetic field is applied since the individual spins are randomly oriented and cancel out when added together throughout the population, as shown in figure 6.2(*a*). In the presence of an applied magnetic field oriented in the positive z (longitudinal) direction, the majority (albeit a rather small majority) of the population of spins is oriented in the direction of the applied field, as shown in figure 6.2(*b*).

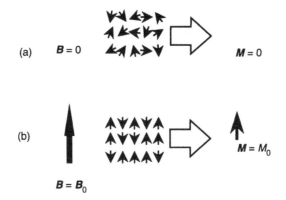

Figure 6.2. In the absence of an external magnetic field ($B = 0$), the individual spins in a volume are randomly oriented, and produce no net magnetic moment (*a*). When an external magnetic field is applied ($B = B_0$), the spins in the volume may align themselves either parallel or antiparallel to the external magnetic field. A small majority of the spins are in the parallel orientation, as determined by the Boltzmann relation, which results in a net magnetization of M_0, aligned with the external field.

The relative quantities of spins in either state is given by the Boltzmann distribution

$$\frac{n^-}{n^+} = e^{\frac{\Delta E}{kT}} = e^{\frac{\gamma \hbar B_0}{kT}} \tag{6.2}$$

where $k = 1.38 \times 10^{-23}$ J K^{-1} (Boltzmann's constant), T = temperature in kelvin, n^- = number of spins oriented parallel to the magnetic field (+z direction), n^+ = number of spins oriented antiparallel to the magnetic field (−z direction), and $\hbar = 6.63 \times 10^{-34}$ J s (Planck's constant). Even in a strong magnetic field, the number of spins in either orientation is approximately equal. However, it is the excess of spins in the parallel state ($n^- - n^+$) which gives rise to the net magnetization, the recordable quantity in MR, conceptualized as a vector sum of individual spins.

6.1.2 Excitation

In the previous section the precession of individual spins was introduced, as was net magnetization. The net magnetization vector also precesses if the precession of individual spins that comprise the net magnetization is synchronized by a process called excitation.

Receiver coils for detecting magnetic resonance signals may detect only the oscillating (due to precession) transverse component of net magnetization—the longitudinal component of magnetization is not detected. To produce a signal that may be recorded, the purely longitudinal, *relaxed* net magnetization must be *excited* by tipping it into the transverse plane. In the presence of the static magnetic field B_0 oriented in the positive z direction, a radio-frequency (RF) magnetic field (B_1) spinning at the resonant frequency is applied in the transverse plane. The resulting torque (using the right-hand rule, $M \times B$) exerts a torque on the net magnetization, which causes the net magnetization to tip toward the transverse plane.

To more clearly illustrate excitation, a rotating frame of reference may be used. Figure 6.3 shows the same excitation event in both the stationary frame (*a*) and the rotating frame (*b*). In the stationary frame (x, y, z) the oscillating B_1 field causes the tip of the net magnetization vector to trace out a spiraling dome pattern as it is tipped away from the z axis until it meets the x–y plane. In the rotating frame (x', y', z), in which the x' and y' axes spin at the Larmor frequency, the oscillating B_1 field is depicted as nonvarying, and the tip of the net magnetization vector simply traces an arc to the y' axis.

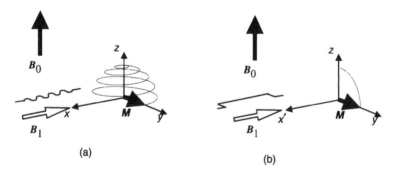

Figure 6.3. An excitation may be represented in either the stationary frame (*a*) or the rotating frame (*b*). With static magnetic field B_0 oriented in the positive z direction and the net magnetization initially oriented along the z axis, an RF pulse (B_1) tips the net magnetization from its relaxed position in the longitudinal direction into the transverse plane.

RF energy is applied for a short amount of time (on the order of several milliseconds). An RF pulse is characterized by the tip angle, α, the angle to which the longitudinal magnetization is tipped. In the example shown in figure 6.3, the RF field B_1 tips the magnetization 90°, and is referred to as a 90° pulse.

6.1.3 Relaxation

After excitation, the net magnetization precesses in the transverse (x–y) plane producing a detectable signal. This signal eventually decays due to two separate effects, which are known by their time constants T_1 and T_2.

Longitudinal relaxation, T_1, is the time required for the net magnetizaton to relax from the transverse plane back to its original position in the longitudinal direction. The precessing transverse magnetization, while experiencing T_1

relaxation, also decays due to a number of other factors, which contribute to what is known as transverse, or T_2, relaxation. T_2 relaxation is due to interactions between individual spins.

Contrast in MR images is possible because different tissues have different T_1 and T_2 relaxation rates. Table 6.1 shows the relaxation rates of several different tissues.

Table 6.1. Relaxation rates for several types of human tissue in a 1 T magnetic field.

	T_1 (ms)	T_2 (ms)
White matter	400	90
Gray matter	500	100
Fat	180	90
Cerebrospinal fluid	2000	300

Transverse and longitudinal relaxation follow the exponential relation shown in equations (6.3) and (6.4):

$$M_z(t) = M_0(1 - e^{-t/T_1})$$

(6.3)

$$M_{xy}(t) = M_{xy}(t=0)e^{-t/T_2}$$

(6.4)

where M_z and M_{xy} are the longitudinal and transverse components of the net magnetization, respectively.

6.2 IMAGING PRINCIPLES

With an understanding of how to excite and record the signal from a single small volume element in tissue, we now describe methods for producing an MR image from a larger volume of tissue made of many such small volume elements. The following sections describe the concepts of selective excitation and spatial encoding, and their use in an example pulse sequence.

6.2.1 Selective excitation

In order to image a particular volume of tissue, only a small slab of tissue is excited. As an example, we describe the task of selectively exciting only a thin slice of tissue in the z direction.

We know from the Larmor relation that spins precess at a frequency that is proportional to the strength of the static magnetic field. If all of the spins within the entire volume were all experiencing the same static magnetic field strength, all spins would precess at the same resonant frequency, and would all be excited by a single RF excitation pulse. However, we may apply a static magnetic field gradient in the z direction, G_z, where the magnetic field in the volume has a linear dependence on z, $B = B_0 k + G_z k$, where k is the unit vector in the z direction. In this case, only a thin slice of spins resonate at the frequency of the applied RF field, and therefore only a thin slice of tissue in the entire volume is excited.

6.2.2 Spatial encoding

After a single slab in the z direction is excited, a signal may be received from the entire slab, but we still are not able to resolve the x and y position of the recorded signal. In order to determine the x and y positions from the MR signal, RF gradients in x and y are applied after excitation in order to encode position in the echoes. Gradient fields, which cause a linear dependence of precessional frequency in either the x or y direction, cause the signals to contain information about the x and y position. This positional information is decoded after the acquisition of MR data, during the process of image reconstruction. Spatial encoding is accomplished by phase encoding (applied before the data acquisition) and frequency encoding (applied during data acquisition).

6.2.3 Pulse sequences

A pulse sequence describes when and how the RF pulses, gradients and acquisition hardware are activated. Pulse sequences are designed according to speed requirements and according to the types of information that is desired in the image. This section introduces the process of producing an echo, and the concept of k-space.

6.2.3.1 Echo production. In order to glean information (T_1 and T_2 relaxation information, for instance) from a MR signal, echoes are produced at a particular time after an excitation. The simplest method of producing echoes is called spin echo.

By tipping the net magnetization into the transverse plane by a 90° pulse applied along the x' axis, the net magnetization is allowed then to dephase, a process that causes the magnitude of the net magnetization to decrease due to individual spins spinning at slightly different frequencies and eventually possessing different phases. Since the magnitude of the signal has diminished, in order to force a detectable signal to be produced, a 180° pulse is then applied, and the ordering of the dispersed spins is reversed. As the spins continue to precess, the magnetization refocuses the individual vectors back into an aligned net magnetization vector. This realigned net magnetization creates a large signal, an echo, which may be detected. Figure 6.4 describes echo production.

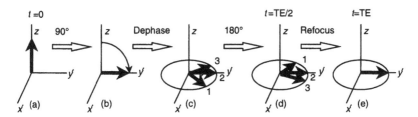

Figure 6.4. To produce a refocused echo, the net magnetization (*a*) is first tipped into the transverse plane by a 90° pulse (*b*). As the transverse magnetization precesses, T_2 relaxation causes dephasing of the original magnetization into three separate components (*c*), with vector 1 precessing at the fastest rate, and vector 3 precessing at the slowest rate. At time *TE*/2, a 180° pulse flips the vectors around about the y' axis (*d*), which puts the fastest moving vector (1) behind the others, and the slowest vector (3) ahead of the others. At time *TE*, the three vectors refocus into a single vector again and create an echo.

A useful analogy for the process of spin-echo production involves a running race. Three runners are told to run toward a given point at a start time, $t = 0$, and each runner runs his/her fastest at a constant, but different, rate (different precessional frequencies). At a given time, $t = T$, the runners are in different positions (dephased), and the runners are told to immediately reverse their direction and return to the starting point (180° pulse). The runners, since they still run at their inherent speeds, return to the starting point at precisely the same time, $t = 2T$ (echo).

6.2.3.2 Pulse sequence diagram. Figure 6.5 shows an example of a spin-echo pulse sequence diagram. Here, the applied RF pulses whose roles in echo production are depicted in vector form in figure 6.4 are shown along with the slice selection (G_z) and spatial encoding (G_y and G_x) gradients, the timing of data acquisition (D/A), and the echo signal which is recorded at time TE.

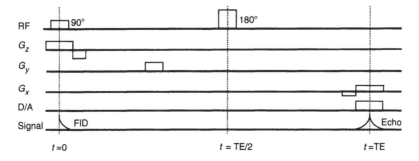

Figure 6.5. A spin-echo pulse sequence applies a 180° pulse at time TE/2, and causes an echo to occur at time TE. This pulse sequence diagram shows the application of slice selection (G_z), phase encoding (G_y) and frequency encoding (G_x) gradients, RF pulses, and time of echo data acquisition with the D/A converter.

6.2.3.3 k-space and Fourier reconstruction. Data acquired by an MRI scanner are collected in *k*-space, which may be thought of as 2D Fourier space. In a spin echo pulse sequence, each horizontal line of *k*-space is the data acquired from a single echo, each encoded differently with spatial information. In order to reconstruct an image from *k*-space data, Fourier reconstruction is most commonly employed. Figure 6.6 shows an illustration of *k*-space, each horizontal line of which is an echo, and a reconstructed image, which is reconstructed by taking the Fourier transform of *k*-space.

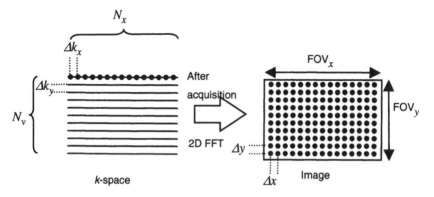

Figure 6.6. *k*-space, made up of digitally sampled individual echoes (horizontal lines) encoded with spatial information, is reconstructed by the 2D fast Fourier transform (FFT) to produce an MR image. Parameters influencing image quality (field of view, spatial resolution, *k*-space sampling resolution, number of samples) are indicated (section 6.3).

6.3 IMAGE QUALITY

Image quality in MR images may be described, in part, on the basis of spatial resolution and signal-to-noise ratio (SNR). Spatial resolution is determined by the following relations which show the one-dimensional case in the *x* direction:

$$\text{FOV}_x = \frac{1}{\Delta k_x} \qquad (6.5)$$

$$\Delta k_x = \gamma G_x \Delta t \qquad (6.6)$$

$$\Delta x = \frac{\text{FOV}_x}{N_x} \qquad (6.7)$$

where FOV_x is the field of view in the *x* direction in cm, Δk_x is the sample resolution in *k*-space in the *x* direction in cm^{-1}, γ is the gyromagnetic ratio (MHz T^{-1}), G_x is the *x* gradient strength (T cm^{-1}), Δt is the sample period (s), Δx is the spatial resolution (cm), and N_x is the number of samples of k-space in the x direction.

The SNR is affected by the volume of image voxels and the total time during which echoes are acquired as shown by the following relation:

$$\text{SNR} \propto V_{\text{voxel}} \times \sqrt{T_{\text{acq}}} \qquad (6.8)$$

where V_{voxel} is the volume of a voxel (cm^3), and T_{acq} is the total time of data acquisition (s). In order to improve the SNR, repetition of scans is often employed to increase acquisition time, and the results are averaged. The number of repeat scans, NEX, increases the SNR by a factor of (NEX)$^{1/2}$.

6.4 MR ANGIOGRAPHY

The imaging of blood vessels with MR is clinically well established today. MR angiography techniques are divided into two main groups by whether or not contrast agents are injected into a patient. The following sections discuss both types, contrast-enhanced methods, and noncontrast-enhanced methods, which include time-of-flight and phase contrast.

6.4.1 Noncontrast-enhanced methods

MR has the ability to quantify and image the flow of blood, and it is this unique ability that is used to image arteries without the use of contrast agents. Phase contrast and spin-tagging methods are the two major classes of noncontrast-enhanced MR angiography methods.

6.4.1.1 Phase contrast. Pulse sequences may be designed to encode velocity information into MR data. The use of flow-encoding gradients allows the creation of images whose intensity is mapped to the velocity of flowing blood. These special gradients cause moving blood to accumulate a phase difference as compared to stationary blood. The disadvantage of phase-contrast methods is that scan times can be lengthy, and they can be subject to dynamic range problems.

6.4.1.2 Spin-tagging methods. Spin-tagging methods use saturation pulses, which reduce the signal in the saturated area, in order to determine flow of blood. Time-of-flight (TOF) methods saturate the region of interest before data acquisition. As unsaturated blood flows into the region of interest, where the stationary tissue is still saturated, the signal in the region of interest increases. The increase in signal is recognized as being due to the in-flowing blood, and images are created that reflect this information.

Figure 6.7 shows an image produced by TOF. The image shown is a MIP (maximum intensity pixel) image, whereby the maximum intensity pixel encountered along rays from a particular projection angle throughout the 3D volume of data is displayed. Other MIP images showing different orientations of the anatomy may be produced from the same set of data, simply by producing the MIP by projecting at a different angle.

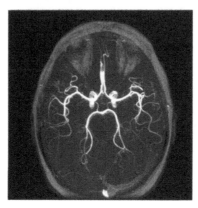

Figure 6.7. Time-of-flight image of the Circle of Willis, an arterial structure in the brain. The image is a maximum intensity pixel image produced from a 3D volume of data (courtesy of Frank Korosec).

Another type of technique, known as the Black Blood technique, does the opposite of the time-of-flight technique. Instead of saturating the region of interest, the blood flowing into the region of interest is saturated. In this way the vessels may be visualized due to the contrast between the blood (low signal) and the surrounding stationary tissue (high signal). The advantage of the Black Blood technique is that the sequences used are relatively fast. However, if blood flow is very slow, this method does not provide any signal. In addition, in regions of complex blood flow, intravoxel dephasing creates artifacts.

6.4.2 Contrast-enhanced MR angiography

Paramagnetic contrast agents shorten the T_1 relaxation time of blood. By injecting these agents into the blood, high-contrast images of vessels may be obtained. Gadolinium, one such contrast agent, is commonly used clinically.

As with X-ray-based angiography, timing of acquisition is critical in contrast-enhanced MR angiography. Arteries are usually the desired vessels to be viewed. Data acquisition is therefore required when Gd injected into the arteries is present only in the arteries. After a short amount of time, the Gd flows into the venous circulation and enhances the signal in the veins as well. In order to obtain images of the arteries without venous enhancement, acquisition must occur between the Gd injection and the time of venous enhancement. If the injection occurs too soon, severe artifacts can ruin the image quality.

Since the timing of venous enhancement is a function of patients' physiological parameters, a low-dose timing scan is often used to estimate the optimal time delay for acquisition. In order to facilitate the task, power injectors, which provide a well-controlled fast injection of the contrast agent bolus, are used.

6.5 DIFFUSION-WEIGHTED AND FUNCTIONAL MRI

Two relatively new applications of MRI that are used clinically are diffusion-weighted and functional MRI. In contrast to other modalities that provide similar information, MRI has the principal advantage that no ionizing radiation is used.

Functional MRI (fMRI) refers most commonly to studies of brain activation. Images in MRI are acquired while a patient is either performing a mental or motor task or being exposed to certain sensory stimuli. The areas of the brain that are active appear as bright areas in the functional image. These functional images are fused with anatomical images to show precisely which areas of the brain are activated. The magnetic properties of oxygenated and deoxygenated blood differ, and these differences serve as the basis for fMRI. In a technique known as BOLD (blood oxygen level dependent contrast), regional cerebral blood flow (rCBF) may be assessed (Rosen *et al* 1998). Figure 6.8(*a*) shows an image produced by fusing an anatomical image with a functional image. The image highlights a region of the brain that is activated when the subject imagines doing a gymnastics routine. The plot in Figure 6.8(*b*) of pixel intensity in the activated region as a function of time shows elevated magnitude while the mental task is being performed.

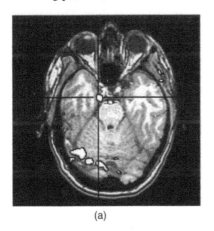
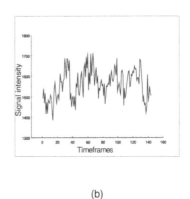

(a)   (b)

Figure 6.8. (*a*) An image produced by fusing a fMRI scan with an anatomical MR scan shows a region of the brain (indicated by crosshairs) that is activated when the subject is imagining herself doing a gymnastics routine. (*b*) shows the signal intensity of the pixels in the indicated region while the subject is either resting (low signal) or performing the mental task (increased signal) (courtesy of Dietmar Cordes).

The ability to detect stroke at a very early stage is of tremendous clinical importance. Conventional MR images allow detection of ischemic events after 6 h or more. Diffusion-weighted imaging allows detection of acute stroke damage 1–2 h after the onset of stroke. In diffusion MR, bipolar gradient pulses are used to make the pulse sequence sensitive to diffusion, which is the random translational motion of water molecules. The gradient pulses magnetically label the spins carried by diffusing molecules, while static spins are unaffected. After a stroke, damaged brain tissue has a higher level of signal in a diffusion-weighted image (Le Bihan 1997).

6.6 MR SPECTROSCOPIC IMAGING

Before MR found its way into the world of imaging, magnetic resonance enjoyed extensive utility in the analysis of the chemical makeup of chemical species in solution in MR spectroscopy. In MR spectroscopic (MRS) imaging, MR imaging and spectral analysis are combined using special pulse sequences to produce images depicting the volume distribution of particular chemical species *in vivo*. Each voxel in an MRS image may be thought of as a spectrum showing peaks that correspond to particular molecules. MRS has received little attention in the clinical radiology community. However, the application has useful and interesting aspects, and it is completely noninvasive. Two common types of MRS are those that involve phosphorus and proton species.

Phosphorus has been used clinically to study changes in high-energy metabolism. The relative levels of adenosine triphosphate (ATP), the major energy carrier in the body, and of other phosphorylated compounds are indicative of the dynamics of anaerobic or aerobic metabolism. The position of the P_i peak in the phosphorus spectrum shifts with cellular pH. The phosphorus spectrum may indicate the relative health of tissue after ischemia. Phosphorus MRS has been used to study stroke, heart disease, muscle disease and tumor metabolism (Sorenson 1998).

MR proton spectroscopy is of particular interest for clinical applications, since existing clinical scanners may be used. In brain tumors, the relative levels of particular molecules in the brain change, which may be used for diagnosis. In addition, studies report the use of MR proton spectroscopy for the assessment of degenerative disorders in the elderly and children, seizure foci, liver disorders, and other brain disorders (Castillo *et al* 1996).

PROBLEMS

6.1   Calculate the precession frequency for a 2 T MRI scanner.

6.2   Sketch a vector diagram of a 45° RF pulse in both the stationary and rotating frames of reference.

6.3   Distinguish between and draw relaxation curves for T_1 and T_2 relaxations.

6.4   For proton MR imaging, we would like an FOV in the x direction of 20 mm. If our sample period is 8 ms, what is the required gradient strength to achieve this FOV?

6.5   Calculate when a 180° pulse must be applied for a spin echo to occur 10 ms after a 90° RF pulse.

6.6   Explain one way to improve the SNR by a factor of 4.

6.7   Describe the concept of spin-tagging methods in MR angiography.

REFERENCES

Castillo M, Kwock L and Mukherji S 1996 Clinical applications of proton MR spectroscopy *Am. J. Neuroradiol.* **17** 1–15

Gould S W T 1997 The interventional magnetic resonance unit—the minimal access operating theatre of the future? *Br. J. Radiol.* **70** S89–97

Lamb G M 1997 Interventional magnetic resonance imaging *Br. J. Radiol.* **70** S81–8

Lardo A C, McVeigh E R, Jumrussirikul P, Berger R D, Calkins H, Lima J and Halperin H R 2000 Visualization and temporal/spatial characterization of cardiac radiofrequency ablation lesions using magnetic resonance imaging *Circulation* **102** 698–705

Le Bihan D 1997 Diffusion and perfusion with MR imaging *Categorical Course in Physics: The Basic Physics of MR Imaging* ed S J Riederer and M L Wood (Oak Brook, IL: Radiological Society of North America) pp 131–44

Nishimura D G 1996 *Principles of Magnetic Resonance Imaging* (Stanford, CA: Stanford University Press)

Prince M R, Grist T M and Debatin J F 1999 *3D Contrast MR Angiography* (New York: Springer)

Redpath T W 1997 MRI developments in perspective *Br. J. Radiol.* **70** S70–80

Rosen B, Buckner R and Dale A 1998 Event-related functional MRI: past, present, and future *Proc. Natl Acad. Sci.* **95** 773–80

Sorenson J A 1998 *The Physics of MRI* (Madison, WI: Departments of Physics and Radiology, University of Wisconsin–Madison)

Wright G 1997 Magnetic resonance imaging *IEEE Signal Process.* **14** (1) 56–66

CHAPTER 7

BIOMAGNETIC AND BIOELECTRIC IMAGING

Alberto Rodriguez-Rivera

Biomagnetic and bioelectric imaging is a modality that emphasizes the functional information of the studied tissue rather than its anatomy. Human organs like the heart, the brain and even muscles can be characterized and studied using this method. This technique is passive, i.e. noninvasive, therefore it is completely safe. Some of the data used to create these images are electroencephalograms (EEGs), magnetoencephalograms (MEGs), electrocardiograms (ECGs), magnetocardiograms (MCGs) and electromyograms (EMGs), among others. This chapter describes the basics of available functional data used to create bioelectric/biomagnetic images, the actual image generation process and bioeffects.

7.1 BIOELECTROMAGNETISM

Bioelectromagnetism is defined as the study of biological fields, either electric or magnetic. Living organisms, specifically humans, work by means of electricity. This underlying electrical activity creates both the electric and magnetic fields. These fields have been studied for a long time and their complexity is considerable. For a thorough exposition on bioelectric and biomagnetic field theory see Williamson *et al* (1982), Malmivuo and Plonsey (1995) and Gulrajani (1998).

Often biomagnetic and bioelectric imaging is used to study and understand human organs. Most of the major organs in the human body function with some form of electricity, e.g. the heart and the brain. This electrical activity generates electric fields that can be detected outside the body. This bioelectromagnetism reflects the well being of the anatomical structure that created it, which is why the study of these fields is of such importance. This imaging modality provides a visual representation of these fields.

Most of the data used for this modality are collected from an array of sensors placed either noninvasively over the skin or invasively by placing them directly over the tissue in question, e.g. next to the wall of the heart. In general, data acquisition is passive, i.e. the body generates the signals spontaneously and we collect them. However, some data are acquired by externally stimulating the organ or tissue in question. This is the case for magnetic readings of the lungs, liver and the electric stimulation of nerves and muscles. We review here some of the available functional data acquisition techniques.

7.1.1 Electroencephalography

Electroencephalograpy can be defined as the recording of brain electric impulses. These electric impulses can be spontaneous or evoked. Spontaneous electric impulses are often used for the diagnosis of brain diseases such as epilepsy. Evoked responses, on the other hand, are often used to detect electric activity that otherwise would be difficult to obtain spontaneously. This is accomplished by recording the EEG of the subject while performing a specific task, i.e. the response is forced by applying an external stimulus. Usually, the experiment is repeated many times and the data are averaged to increase the signal-to-noise ratio (SNR). The electrodes are placed on the scalp or invasively laid on top of the cerebral cortex.

Hans Berger recorded the first EEG in 1924. Surface potentials are in general faint; the typical amplitude of a surface EEG on humans is about 100 μV. However, at the cortex the typical amplitude of the signal is about 1–2 mV, i.e. around 20 times greater than surface potentials. These cortical potentials can be acquired by placing a grid of electrodes directly over the surface of the cortex. Changes in the potentials are due to the firing of a population of neurons. However, even when the neuron is at rest, there is a potential present. The neuronal resting potential is about –70 mV. The number of neurons in an average human is about 10^{11} cells.

The typical EEG recording uses multiple electrodes on the patient's scalp. The locations of these electrodes are carefully chosen to cover the entire head area. During a regular EEG session the electrodes are positioned in a so-called *10–20 standard*. The number of electrodes for an EEG can vary from just a few electrodes all the way to hundreds. The electric impulses reflect the well being of the neural tissue. It is common clinical practice to read EEG waves to diagnose a myriad of brain dysfunctions.

7.1.2 Magnetoencephalography

Magnetoencephalography can be defined as the detection and recording of weak magnetic fields generated by the brain. The brain's magnetic field is 100 to 1000 times smaller than the heart's magnetic field, and one billion times smaller than the earth's magnetic field. The problem of detecting such a faint signal is of monumental complexity. It was not until the end of the 1960s that technology allowed researchers to consistently detect those fields. Even with today's technology it is necessary to place the detectors inside a magnetically shielded room to have a chance of detecting those faint magnetic fields.

The first MEG was recorded in 1968 with coil detectors while the first MEG using SQUIDs was recorded in 1972. SQUID stands for superconducting quantum interference device. The same electric activity that creates the EEG is the one that creates the MEG. At the beginning it was thought that MEGs and EEGs yielded nonoverlapping information. However, later studies demonstrated that there is much overlapping information between the two modalities. Nevertheless, magnetoencephalography still attracts many people because the magnetic fields are not affected by the skull nearly as much as the electric field. EEGs are still preferred over MEGs for brain studies mainly because the EEG is cost effective: the price of a good EEG machine can be measured in thousands while a good MEG machine is measured in millions. Many facilities are using MEG magnetometers for research purposes.

7.1.3 Electrocardiography

Electrocardiography can be defined as the recording of the electric impulses emanating from the heart. The ECG is noninvasively acquired by placing an array of electrodes over the chest cavity and sometimes around the torso and back. It can also be acquired by placing the electrodes directly over the heart's walls or in the heart's chambers.

Augustus Desire Waller recorded the first human ECG in 1887. It was not until 1913 that Einthoven recorded the ECG using a string galvanometer. This is a very cost-effective way to diagnose many heart conditions, including arrhythmias and myocardial infarctions. However, the ECG has a very poor spacial resolution making it difficult to generate an image from the regular 12-lead ECG. This problem is partially solved by using *body surface potential mapping* (BSPM). BSPM is an array of hundreds of electrodes covering the entire torso and most of the back. However, to place the array is very time consuming and therefore is mostly used for research purposes.

7.1.4 Magnetocardiography

Magnetocardiography can be defined as the recording of the weak magnetic fields generated by the heart muscle. Baule MacFee recorded the first MCG in

1963. The heart's magnetic field is stronger than the brain's magnetic field, but still is far weaker than the earth's magnetic field. As in the case of the MEG, the MCG sensors are placed as close as possible to the patient's thorax because the magnetic fields attenuate rapidly as a function of distance. The MCG provides some information that the ECG cannot provide. Still the MCG is not, in general, used in the clinical world and it remains a research tool mostly because of economics.

7.1.5 Biosusceptometry

Biosuscepometry can be defined as the measurement of evoked magnetic fields generated by magnetic contaminants or tracers inside body organs. These fields are in the range of nT to pT. Some examples of biosusceptometry are magnetopneumography, liver susceptometry and magnetic susceptibility plethysmography.

Magnetopneumography is the study or measurement of the magnetic fields evoked by ferro- and/or ferrimagnetic contaminants or tracers in the lungs. The magnetic fields used for excitation are of the order of 1 mT to 1 T while the evoked magnetic field is on the order of nT. Liver susceptometry is used for quantifying iron levels in the liver. A liver biopsy is the invasive alternative to liver susceptometry. Other modalities such as quantified magnetic resonant imaging (qMRI) are also developing for this application. Finally, magnetic susceptibility plethysmography is mainly used to measure cardiac volume changes. The magnetic fields are of the order of 1–100 pT (Fischer and Heinrich 1992). In general we can say that biosusceptometry is being replaced by other more effective modalities such as sonography, X ray, CT, PET and MRI.

7.2 IMAGE GENERATION

Bioelectric and biomagnetic image generation starts with the acquisition of electric and/or magnetic data from the body. With the functional data, e.g. EEG, MEG, ECG, etc, many different types of algorithm might be used to create the images. The images can be presented on fake body parts such as average heads, torsos or hearts that approximate the real geometry of the tissue in question. On the other hand, 3D models from anatomical scans can be created to overlay the functional data. This is also called multimodal imaging and is discussed in chapter 9.

Figure 7.1 shows that image generation uses two independent processes, one regarding functional data and one for the anatomical data. The functional data are collected and postprocessed to create surface-potential mappings that are overlaid over the results obtained from the anatomical processes. The anatomical process is fed with MRI, PET or CT scans for the construction of the 3D models. Those models, together with the surface-potential maps, conform the bioelectric/biomagnetic image. To better understand the process we now review it with some examples.

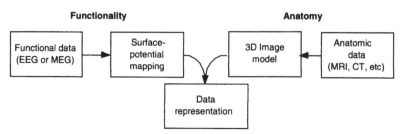

Figure 7.1. Bioelectric and biomagnetic images completely rely on functional data to create their images. The anatomical modeling is optional but very helpful for better visualization.

7.2.1 Heart bioelectric or biomagnetic imaging

Heart bioelectric/biomagnetic imaging is a new technique that has been used mainly for research purposes. The simplest way to create a heart electric image is to have a dense electrode array over the patient's torso. The electrocardiogram is an array of time series representing the electric activity of the heart at some predetermined locations. The locations of the electrodes can be placed into an arbitrary coordinate system so their positions can be displayed in three dimensions. A torso resembling the real one is often used for image visualization. Figure 7.2 shows an example of a bioelectric image overlaid on an average torso (not real).

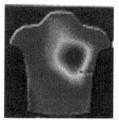

Figure 7.2. A bioelectric image of the heart projected into the torso. One of the advantages of this kind of data representation is the increased number of available data that are presented at one time (courtesy of Brooks and MacLeod 1997).

The simplest way to create the image sequence is when the desired projection surface coincides with the surface where the electrodes are placed. To create an image at time *t*, we define a color scale for all voltage values. We now take every sensor output and map a color to its value. This will yield as many color points as the number of sensors present. To fill the rest of the surface, we simply interpolate to obtain colors everywhere. This is called spatial interpolation. Notice the word spatial, which means that this is done for every time sample. This implies that the images are time dependent. Simple spatial interpolation is very useful and informative, however; it does not help to reduce the amount of information present.

Notice that the contour image is displayed over the surface of the thorax, as shown in figure 7.2. If the contour image is wanted over the surface of the heart a more complex process is necessary. One way is to place electrodes directly over the heart muscle, which reduces the problem back to the simplest example explained above. This is done to patients that are scheduled for open-heart surgery only. After the ECG from the walls of the heart is collected, spatial interpolation is performed and an image is generated. However, the diagnostic advantages of a bioelectric image of the heart do not outweigh the open-heart surgery. Another alternative is to use noninvasive methods, often called *inverse methods*. This method will be discussed later in the chapter.

Spatial interpolation is good because the interpretation of the data is left to a human. Humans have a great ability for temporal and spatial interpolation and edge extraction (Brooks and MacLeod 1997). Algorithms that take advantage of the temporal information are also available. *Isointegral mapping,* for example, integrates every sensor individually and produces a single picture regardless of how many data are available. This has the advantage of great data compression. However, data loss is the price to pay. For an overview of more algorithms read Brooks and MacLeod (1997).

7.2.2 Brain bioelectric or biomagnetic imaging

Brain imaging is one of the common applications for this imaging modality. A bioelectric or biomagnetic study is often used for epilepsy patients that will undergo brain surgery or to understand the behavior of the brain under certain circumstances. Figure 9.11 shows a brain bioelectric image mapped to a head model. This model can be an average version of a human head or it can be realistically created from MR images. The processes to create these head models are relatively simple and are well documented. Creating a realistic model begins with a high-resolution MR scan. With the scan you can create a 3D model. This model should include some reference points that will be used later: usually these are the so-called fiducial points.

The exact 3D location of the electrode array should be recorded together with the fiducial points. Then the electrodes are transformed to the anatomical coordinate system created from the MR scan. Now we have all the information in the same coordinates. We now take every electrode potential and map a color corresponding to its magnitude. To create the image we do spatial interpolation as explained in the previous section. Again, if the intended surface of projection does not match the surface where the electrodes lie, other more complex methods should be used, e.g. inverse methods.

7.2.3 The inverse problem

The *inverse problem* refers to the estimation of electric activity inside the body deduced from surface potentials or magnetic fields. A major obstacle is that there is no unique solution for this problem, i.e. it is ill posed. Moreover, the signals are smoothed and attenuated as they exit the body, which means that most of the data are lost forever. One way to approach this problem is to find the *forward model* and use it to solve the inverse problem. Still we have the uniqueness question, i.e. the number of solutions is infinite. To solve it we have to use *a priori* information such as realistic geometry, conductivity and good data models to decrease the number of solutions.

The forward problem is the mathematical expression that relates sources in a biological volume to a set of sensors. This is true regardless of the modality used, electric or magnetic. The mathematical equations contain information about the conductances of the tissues that the fields will encounter during the travel, as well as geometrical information about the conductor volume. The volume conductor models can be analytically or numerically solved according to the selected geometry. If the geometry is extremely irregular, as in the case of the cerebral cortex and the heart, an MRI or CT scan is used to extract information about this geometry. See chapter 9 for more details on this procedure. This realistic model makes an analytical solution impossible. Finite element methods (FEMs) and boundary element methods (BEMs) are common methods. The FEM can predict potentials at all nodes of the array while the BEM uses only surface equations with fewer degrees of freedom. To obtain volume potentials from the BEM requires a second step. This method has difficulty handling anisotropy.

The inverse problem can be seen as an image restoration problem (Brooks and MacLeod 1997). From an image of the field contour at the surface a picture of the contour inside must be obtained. Mathematically speaking, we can pose this problem as follows. We assume there is a linear function relating an internal electrode array and the surface electrode array called A. Let y be a vector that contains the data of the surface array, and let x be the vector that contains the data of the internal array. These quantities are related by $y = Ax$. The available data are y, the ECG or EEG, and A, the lead field matrices that can be mathematically computed. Moreover, we assume that the internal data are unknown. Then, the projection or regularization problem can be expressed by

$$\arg_x \min \left(\|y - Ax\|^2 + \lambda \|x\|^2 \right) \qquad (7.1)$$

where λ is called the regularization parameter. There are some issues that make this problem particularly difficult. In image processing, regularization is done on a 2D image, often as a regularly sampled squared image. This problem, on the other hand, is not sampled regularly (spatially) and the surface to project the inverse solution is different from the one we started with. These problems are partially solved when anatomic images are introduced and a unified coordinate system is defined. Section 9.3.3 shows an example of the results of this procedure.

7.2.4 Spatial and temporal resolution

The accuracy of the resultant pictures depends on spatial as well as temporal resolution. Spatial resolution is dependent on the number of electrodes available while temporal resolution is dependent on the sampling rate. For electrocardiography, for example, an array of electrodes with interdistances of 10–20 mm is often used for imaging applications. For internal arrays, on the walls of the heart, a grid of 5–10 mm is usual. The spatial resolution is closely tied to storage space and availability of resources. If the number of sensors is very high, the equipment cost will be higher and the amount of data storage required will also increase.

Temporal resolution also has a similar problem. If the sampling rate is increased the amount of data storage needed will increase accordingly. For heart applications, surface electrodes are usually sampled from 250 to 1000 Hz, while internal arrays, e.g. in the walls of the heart, are sampled from 1000 to 2000 Hz.

7.3 BIOEFFECTS

There are no negative biological effects associated with bioelectric imaging. The necessary data to create the image are passively acquired by using electrodes. For biomagnetic images, biological effects are not found if passive data acquisition is performed. However, in the case of liver susceptometry and magnetopneumography, where external magnetic fields are applied, some bioeffects may occur. The mechanism of how these bioeffects occur is very similar to the bioeffects of magnetic resonant imaging discussed in chapter 6.

PROBLEMS

7.1 Describe how to obtain an image using the EEG.
7.2 Describe how to obtain an image using the MEG.
7.3 Describe how to obtain an image using the ECG.
7.4 Describe how to obtain an image using the MCG.
7.5 Describe how the inverse problem can be solved to obtain an image.

REFERENCES

Brooks D H and MacLeod R S 1997 Electrical imaging of the heart *IEEE Signal Process.* **14** (1) 24–42

Fischer R and Heinrich H C 1992 Biosusceptometry current status of clinical diagnostics and biomagnetic research *Biomagnetism: Clinical Aspects* ed M Hoke, S N Erne, Y C Okada and G L Romani (New York: Elsevier)

Gulrajani R M 1998 *Bielectricity and Biomagnetism* (NewYork: John Wiley)

Malmivuo J and Plonsey R 1995 *Bioelectromagnetism. Principles and Applications of Bioelectric and Biomagnetic Fields* (London: Oxford University Press)

Nunez P L 1995 *Neocortical Dynamics and Human EEG Rhythms* (London: Oxford University Press)

Regan D 1989 *Human Brain Electrophysiology: Evoked Potentials and Evoked Magnetic Fields in Science and Medicine* (New York: Elsevier)

Williamson S J, Romani G, Kaufman L and Modena I 1983 Biomagnetism: an interdisciplinary approach *Proc. NATO Advanced Study Institute on Biomagnetism* (New York: Plenum Press)

CHAPTER 8

ULTRASOUND

Alberto Rodrìguez-Rivera

The study of medical imaging would not be complete without discussing ultrasound. Even though it does not provide the anatomical resolution of other modalities, it is relatively inexpensive and safe, making it one of the most popular imaging techniques. Ultrasound may be used for the minimally invasive diagnosis of illnesses of the vascular system, thyroid and parathyroid glands, breast lumps and fetal development assessment, just to mention a few. Even though therapy is another area where ultrasonic systems are used we focus our attention on diagnostic imaging. We provide a basic introduction to the physical principles, instrumentation and bioeffects of ultrasonic imaging, as well as Doppler ultrasound.

8.1 PHYSICAL PRINCIPLES OF ULTRASOUND

Ultrasound, as the name suggests, uses sound waves to generate images. The frequencies used for ultrasound range from 2 to 10 MHz. Sound waves are partially reflected at tissue boundaries, and these reflections are called *echoes*. The amount of reflection is dependent on the tissue density differences as well as the incidence angle of the wavefront. The time elapsed since the signal generation and its echo is used to compute the distance of the object. Information about the shape and angle of arrival is used to estimate the shape and location of the object. Figure 8.1 shows a general block diagram of an ultrasound system.

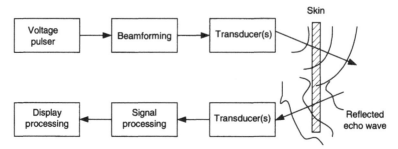

Figure 8.1. An ultrasound system transmits ultrasound through the skin and detects reflected echoes.

8.1.1 Sound waves in sonography

Sound waves propagate through the air, or any other medium, by exciting molecules and transmitting the energy to adjacent molecules, very much like balls on a pool table pushing each other. Waves that vibrate perpendicular to the traveling direction are called *transverse* waves and waves that vibrate parallel to the traveling direction are called *longitudinal* waves. Longitudinal waves create small disturbances on the molecules as they travel into the body, creating zones where the molecules are highly concentrated and others where they are sparse.

8.1.2 Speed, wavelength and frequency

The period (T) of a sound wave, measured in seconds, is the time for one cycle to be completed, in other words the time for a wave pattern to repeat itself. The wavelength (λ), measured in meters, is the length of one cycle. The frequency (f), measured in Hz or ($1/s$), is the number of times the wave pattern repeats itself over one second. The propagation speed (v), measured in m s^{-1}, is computed as the wavelength multiplied by the frequency, $v = \lambda f$ m s^{-1}.

In human tissue the range of the speed of sound is between 1450 and 1600 m s^{-1}. Density and compressibility of the medium determine the propagation speed. The propagation speed is approximately inversely proportional to the density and the compressibility of the medium.

8.1.3 Sound intensity

In addition to speed, wavelength and frequency of sound, we also consider loudness or intensity. Intensity (I), measured in watts per meter squared (W m^{-2}), is defined as the average rate per unit area at which the wave transmits energy, as expressed in equation (8.1). Watts is a measurement of power and can also be defined as joules per second (J s^{-1}). Mathematically, sound intensity is given by

$$I = \frac{1}{2}\rho v\omega^2 a^2 \tag{8.1}$$

where ρ is the volume density of the tissue, $\omega = 2\pi f$ and a is the amplitude of the wave. The wave amplitude can be defined as the maximum displacement of molecules due to the wavefront impact. This wave or displacement amplitude at the human ear ranges from a maximum of tolerable loudness of 10^{-5} m to the faintest sound of 10^{-11} m.

Due to the wide range of amplitudes that we study, sometimes it is better to work with a logarithmic scale. This scale is based on the ratio of intensities between two waves and is named the *bel* scale (or the sound level), in recognition of the work of Alexander Graham Bell. Assuming that the intensity of the reference wave is I_0 and the intensity of the wave in question is I, the intensity under the bel scale is given by

$$10 \log_{10} \frac{I}{I_0} \ \text{(dB)}.$$ (8.2)

The bel system is often measured in decibels in the literature rather than bels, where 1 bel = 10 dB.

8.1.4 Sound behavior and its interaction with objects

As sound waves travel through human tissue they experience attenuation. Spreading is an attenuation mechanism that depends on distance. Spreading refers to the distribution of energy over an area, the larger the area the smaller the energy at a given location. A rule of thumb is that 1 dB of energy is lost for each centimeter of tissue penetration per MHz, e.g. a 5 MHz signal loses 5 dB of intensity for every centimeter of depth.

Reflection and transmission are other mechanisms of attenuation, as shown in figure 8.2. The amount of reflection depends on the angle of incidence (θ) of the sound wave. If the sound wave hits a dense surface at an incident angle of 90° most of the signal is reflected, just as light reflects from a mirror. These flat surfaces are called *specular reflectors*. Nevertheless, if the sound wave bounces to a place where there is no receiver, the signal is lost. Transmission occurs in structures with similar density where the sound wave completely penetrates the medium and nothing is reflected.

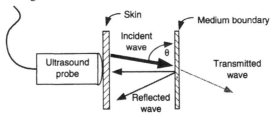

Figure 8.2. Reflection and transmission are two types of loss in ultrasound. If the incident wave hits a tissue boundary at a wrong angle the reflection or echo may not reach the ultrasound probe, resulting in a loss. The wave energy that is not reflected is transmitted and, if there are no other boundaries, this signal never gets back to the source.

Attenuation due to scattering occurs on rough boundaries, as shown in figure 8.3. These types of surface are hard to detect due to the poor reflection. Heat transfer is another type of attenuation. Sound waves are pressure waves that excite molecules in the air, exchanging energy as they collide. This excitation is partially dissipated as heat.

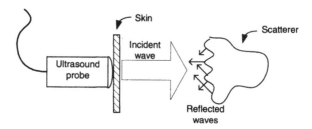

Figure 8.3. Attenuation due to scattering occurs at boundaries of objects of irregular shape.

As the sound wave continues traveling through the medium the portion that is not attenuated or lost continues its journey. Tissue impedance is computed by

$$Z = \rho \times v \tag{8.3}$$

where ρ is the volume density and v is the speed of sound within the tissue. The amount of reflected energy is computed by

$$R = \left(\frac{Z_2 - Z_1}{Z_2 + Z_1} \right)^2 \tag{8.4}$$

where Z_1 is the impedance at the incident boundary and Z_2 is the impedance at the transmitting boundary. The traveling path and speed behave in a predictable way as explained by Snell's Law:

$$v_i = v_t \frac{\sin(\theta_i)}{\sin(\theta_t)} \tag{8.5}$$

where v_i and θ_i are the incident velocity and angle from the normal, and v_t and θ_t are the transmitted velocity and angle from the normal.

The equation relates the speeds and angles of incidence at a boundary. Knowing the densities and compressibility characteristics of the medium we can compute the speed of propagation and with Snell's Law predict the transmitted angle of refraction.

8.2 TRANSDUCERS

A transducer is a device that changes one form of energy to another: in ultrasound, voltage is changed to sound and vice versa. Some transducers are naturally piezoelectric, such as quartz, i.e. they transform pressure to electricity and vice versa. Nonpiezoelectric materials can become piezoelectric by exposing them to powerful electric fields while they cool from high temperatures.

8.2.1 Transducer resonant frequency

Transducers have a preferred frequency of operation, called the *resonant frequency*. The resonant frequency can be computed as

$$f_r = \frac{v}{\lambda} \tag{8.6}$$

where v is the propagation speed in the crystal and λ is the wavelength. The wavelength in turn can be computed as $\lambda = 2D$, where D is the crystal thickness or diameter if it is circular.

The quality factor is

$$Q = \frac{f_r}{BW} \tag{8.7}$$

where BW is the bandwidth. BW is computed as the difference between the frequencies above and below the resonant frequency where the amplitude is half the resonant amplitude, as shown in figure 8.4. This measure of quality is helpful when selecting the optimal transducer for a specific application.

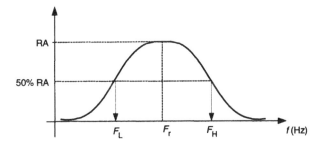

Figure 8.4. To compute the bandwidth, first locate the resonant frequency (F_r). Then find the point that corresponds to the resonant amplitude (RA). Compute 50% of the amplitude and locate the frequencies that correspond to it. The frequency that is higher than the resonant frequency is called F_H and lower is called F_L, the difference of these frequencies is called the bandwidth.

8.2.2 Transducer assembly head

An ultrasound machine is basically a computer with a set of probes. The transducer assembly, located inside the probe housing, is composed of three layers: a backing layer, a piezoelectric crystal and a matching layer, as shown in figure 8.5. When the piezoelectric material is excited it expands and/or contracts depending on the polarity of the pulse, generating a longitudinal pressure wave (sound wave) or vice versa. The amount of deformation is directly proportional to the voltage applied.

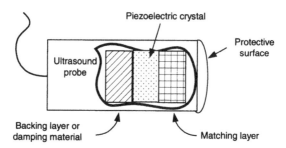

Figure 8.5. In the ultrasound probe, the backing layer damps the short pulses, the piezoelectric crystal transforms voltages into sound waves and the matching layer matches impedance.

The backing layer is used as a dampening device. For diagnostic ultrasound imaging short sound pulses are used rather than continuous signals. Short voltage pulses do not yield short sound waves, therefore a dampening device is required, as shown in figure 8.6. The materials used for this layer are usually tungsten and either rubber or plastic held together with epoxy.

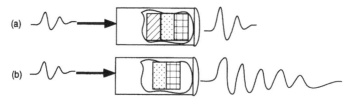

Figure 8.6. (*a*) A transducer assembly head with a backing layer attenuates the sound output to produce short pulses. (*b*) A transducer without a backing layer generates longer pulses, which yield less axial resolution.

The piezoelectric crystal is central to the ultrasound system. Some of the materials used to create the crystal are different types of zirconate titanate, which is a ceramic. This crystal has inherent impedance that is by no means optimal for power transmission. To overcome this transfer problem a matching layer is needed.

The matching layer is an impedance equalizer. This enhances the sound wave transmission by allowing maximum power transfer. This layer's impedance is a value between the crystal's impedance and that of the skin. Finally, a layer of gel is used to avoid air between the skin and transducer assembly contact area. If the gel is not used a thin air layer might be present in the contact area creating a large impedance boundary, which causes the sound wave to be completely reflected.

8.2.3 Types of transducer assembly head

Many types of ultrasound piezoelectric crystal exist today. There are single or multiple element assemblies that are either mechanical or electronic. Single-element mechanical assembly heads oscillate back and forth in order to scan the desired area, as shown in figure 8.7(a). Multiple rotary assembly heads revolve around an axis to achieve scanning, figure 8.7(b). Electronic assembly heads rely on linear arrays of piezoelectric crystals, figure 8.7(c). Timing the order in which the linear array of crystals is excited creates synthetic scanning.

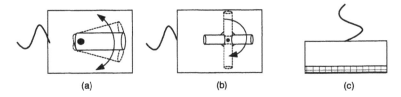

(a)   (b)   (c)

Figure 8.7. (a) A mechanical single piston transducer has to oscillate rapidly to scan the required area. (b) The mechanical rotatory array scans an area by rapid revolutions. (c) The electronic transducer array has many small components. To scan an area it excites a select group of transducer elements.

8.2.4 Sound beams

As the sound beam penetrates the tissue, spatial and temporal attenuation occur. At a fixed position in space, a measurement of the intensity at time t and then at time $t + \Delta t$ yields a nonzero difference. Therefore there is a temporal dependence. Similarly, at a fixed time, the intensity at two different points in space yields a nonzero difference. Therefore there is a spatial dependence.

Sound beams behave differently depending on how far they are from the probe. The *Fresnel zone* is where the beams are parallel to the probe and there is little or no divergence of the beam, this is also known as the *near zone*, as shown in figure 8.8(a). The *Fraunhofer zone* is where the beam starts to diverge; this is also called the *far zone*.

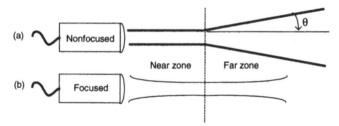

Figure 8.8 (*a*) A nonfocused sound beam propagates ideally in a straight line up to the Fresnel distance, i.e. at the end of the near zone the beam starts to diverge at a certain angle. (*b*) The focused beam improves the transverse resolution by concentrating the sound beam over a smaller area, with the smallest beam width occuring at or before the Fresnel distance. Note that the tip of the probe is not a crystal but a protective surface.

The Fresnel zone is given by

$$F_z = \frac{r_t^2}{\lambda}$$
(8.8)

where r_t is the transducer's radius and λ is the wavelength. The divergence angle in the far zone is given by

$$\theta = \sin^{-1}\left(1.22\frac{\lambda}{d}\right)$$
(8.9)

where d is the diameter of the crystal.

The focused beam also has near and far zones, as shown in figure 8.8(*b*). The idea of the focused beam is to increase transverse resolution, just as for the case of light. The focal point or the point where the beam width is the smallest is less than or equal to the Fresnel distance. Note that the tip of the ultrasound probe does not follow the crystal shape but has a protective surface. There are several ways to generate a focused beam. A concave piezoelectric and a concave sound lens in front of the piezoelectric generate a focused sound beam, as shown in figures 8.9(*a*) and (*b*). A linear phased array can also generate focused beams by firing the outermost elements first, as shown in figure 8.10.

Figure 8.9. (*a*) A concave piezoelectric crystal generates a focused sound beam. (*b*) A concave sound lens in front of the piezoelectric crystal generates a focused sound beam.

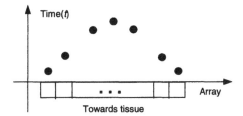

Figure 8.10. A linear phased array can be used to create focused sound beams by firing the outer elements of the array first. The vertical axis represents time, the dots represent the instants in time when the elements right below them were activated. The central element is the last to be fired.

8.2.5 Transducer beamforming

The word beamforming implies the steering or directional control of the sound beam. The ultrasound probes with mechanical systems steer the sound beam by physically pointing the piezoelectric crystal. The electronic ultrasound probes do it by firing array elements in a predetermined way. The addition of many waves will determine the direction of propagation, as shown in figure 8.11.

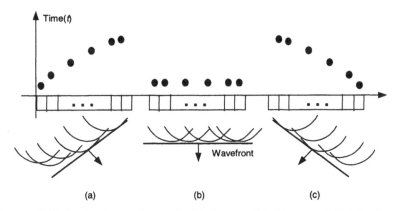

Figure 8.11. (*a*) Steering the wavefront to the right is accomplished by sequentially firing the array elements from left to right. (*b*) Firing all the elements simultaneously generates a wavefront that travels parallel to the probe head. (*c*) Firing the array elements from left to right generates a wavefront that travels to the left.

8.3 ULTRASOUND IMAGE GENERATION

Every sound beam generated represents one line of information, as shown in figures 8.12(*a*) and (*b*). The penetration of the sound beam depends on how much time elapses waiting for the echoes. Mechanical transducers, as shown in figure 8.13(*a*), often generate pie images. Rectangular images, as the one shown in figure 8.13(*b*), are usually generated by a linear array. Every line is composed of dots or pixels. Each pixel has a gray scale value chosen according to the magnitude of the received echo. The difference between representable maximum and minimum is called the *dynamic range*. Usually one image or scan is composed of 128 lines. Normally between 24 and 30 scans are presented per second. To increase the number of scans per second we need to pulse faster but the penetration suffers, i.e. only shallow objects can be studied. A trade-off between penetration and screen update is required.

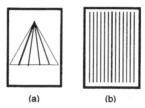

(a)   (b)

Figure 8.12. Ultrasound scans are composed of multiple lines pointing to different locations. Each line is generated by one pulse. (*a*) Pie-shaped ultrasound. (*b*) Rectangular-shaped ultrasound.

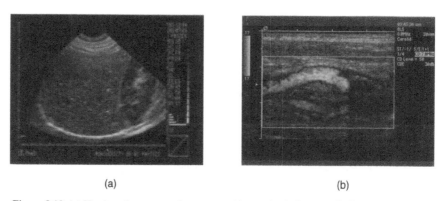

(a)   (b)

Figure 8.13. (*a*) Pie-shaped scans are often generated by mechanical means. (*b*) Rectangular-shaped scans are often generated by linear arrays.

Ultrasound has many types of signal representation. They are called modes. The 'A' mode or amplitude mode determines tissue boundaries by looking at peaks in a time series, i.e. looking at maxima points as a function of time. This mode does not yield a picture and is often used for ophthalmology. The 'B' mode or brightness mode is very popular for diagnostics. The data are represented in the usual gray scale image, as shown in figures 8.13(*a*) and (*b*). The 'M' mode or motion mode allows the user to see how a given pixel evolves in time by plotting the results on a rolling screen, very much like printing on continuous paper.

8.3.1 Ultrasound resolution

Resolution in ultrasound has to do with the ability of the system to distinguish between objects divided by small distances as well as the differentiation of small echo magnitudes. These two cases are often called *detail* and *contrast* resolution respectively (Williamson 1996).

Two objects along the wave's traveling direction, also referred to as the *axial* direction, can be differentiated if the distance between them is at least one-half the pulse length, see figure 8.14. To increase resolution shorter pulses can be used. The size of the piezoelectric crystal also affects resolution. Lateral (transverse) resolution refers to the differentiation of objects distributed in the axis orthogonal to the axial direction, as shown in figure 8.15. Small piezoelectric crystals can generate narrow sound beams. Lateral resolution is also affected by distance. According to Fresnel, the sound beam diverges according to equations (8.8) and (8.9).

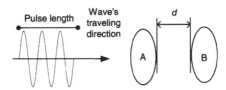

Figure 8.14. If the distance, *d*, between object A and B is less than one-half the pulse length, then the objects are not differentiated. Remember that one pulse yields one line worth of data.

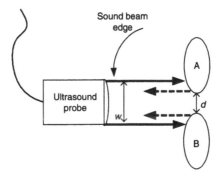

Figure 8.15. The beam width, *w*, determines the lateral resolution. Small beam widths yield better resolution.

Contrast resolution is determined by how many gray levels can be represented in a given system. As the number of bits for contrast increases, the number of gray levels increases exponentially. For example, a one-bit representation for contrast yields only black and white. A two-bit representation yields four levels, and so on.

8.3.2 Artifacts

During the design process of an ultrasound system certain assumptions have to be made in order to maintain a good level of accuracy of the results. Nevertheless, these assumptions are often violated, which creates artifacts. The assumption that the sound beam travels in one direction and the echoes return from the same direction is often violated. Some of the factors that cause this problem are reverberations, multipath echoes, sound leakage, refraction and others.

Reverberations are caused by sound waves trapped between two objects, as shown in figure 8.16. Multiple copies of nonexisting structures are present along the axial direction. Refractive artifacts are created by echoes arriving from sources that are at an angle from the axial direction, as shown in figure 8.17. The structure that is actually at an angle is depicted in the axial direction and deeper than it actually is. Another refractive artifact is known as the mirror image effect. This occurs when sound strikes a structure that is highly reflective, like a mirror, striking a structure that is at an angle from the axial direction. Sound returns to the mirror structure and finds its way back to the scan probe, see figure 8.18. This results in a false deep structure along the axial direction. Multipath artifacts are very similar to the mirror effect artifacts in the sense that deep structures are falsely identified. Multipath begins with a reflective structure that moves the sound beam away from the axial direction striking a second structure that sends the sound beam to the scan probe, see figure 8.18.

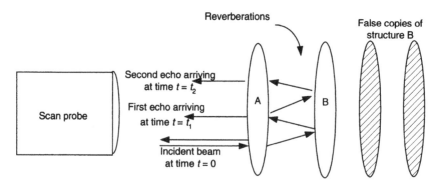

Figure 8.16. Reverberation artifacts are caused by sound beams trapped between two structures. The results will be the identification of structures A and B plus false copies of structure B, also known as phantoms.

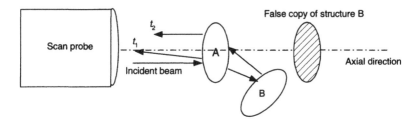

Figure 8.17. Refractive artifacts are caused by the detection of objects that are at an angle from the axial direction. Copies of the B structure will be falsely detected deep along the axial direction.

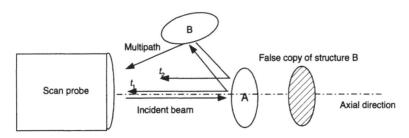

Figure 8.18. The mirror image artifact occurs when a structure 'A' reflects the sound beam to structure 'B' away form the axial direction. Structure 'B' reflects to 'A' and back to the probe. Part of the sound will bounce from 'B' directly to the scan probe, causing a multipath artifact.

Shadowing is caused by a strong reflective structure. When most of the sound is reflected back to the scan probe no information from deeper structures is available, creating a black zone (shadow). Sound leakage artifacts are created by the generation of a side lobe of sound energy coming from the scan probe, as shown in figure 8.19. This is a clear violation of the assumption that the sound beam and the echoes travel parallel to the axial direction.

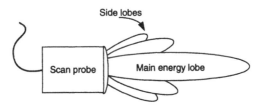

Figure 8.19. Sound energy does not travel completely along the axial direction. Part of the energy travels at an angle from the axial direction. Structures encountered by these side lobes will generate echoes that will be assumed to be along the axial direction.

8.4 DOPPLER ULTRASOUND

Doppler ultrasound is mainly used to estimate motion. The Doppler effect is a very old concept. Christian Andreas Doppler first explained the phenomenon in the 1800s. People often mistakenly attribute the discovery of this phenomenon to Johan Doppler, Christian's father (Kremkau 1995). Animals such as bats and dolphins use this concept for navigation. Sound waves striking surfaces in motion experience a change in frequency, called the Doppler shift. Doppler shift measures relative movement between the source and the studied object. Objects moving toward each other experience an increase in frequency. Objects moving away from each other experience a decrease in frequency.

In medicine we use this phenomenon to measure blood flow. Mathematically speaking the Doppler shift, D, is computed by

$$D = \frac{2f_0v}{m}\cos(\theta) \tag{8.10}$$

where f_0 is the emitter frequency, v the object velocity, m the speed of sound in human tissue (1540 m s^{-1}) and θ the angle between the sound propagation vector and the object velocity vector, see figure 8.20.

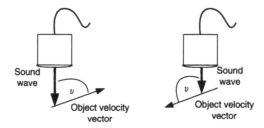

Figure 8.20. To define the angle θ, the sound propagation vector and the moving object velocity vector are needed.

8.4.1 Continuous wave Doppler ultrasound

Continuous wave Doppler (CWD) is simple. The emitted sound wave is continuously generated. Thus there must be two different piezoelectric crystals for emitting signals and receiving echoes. With CWD no depth information can be obtained making this modality good for superficial structures. Notice that the Doppler shift as computed in equation (8.10) does not convey directional information, i.e. if the shift is positive or negative.

Phase V quadrature is used to determine direction (Kremkau 1995). Quadrature is obtained when the received signal is multiplied by a sinusoidal term that is 90° out of phase with a cosinusoidal in phase term. Both terms are modulated versions of the received signal. Both in phase and quadrature terms have the same frequency and phase terms. Hence, if the received frequency is greater than zero the quadrature term is retarded with respect to the in phase term and if the frequency is less than zero the opposite is true.

8.4.2 Pulsed Doppler ultrasound

Pulsed Doppler (PD) systems, as the name suggests, obtain information by sending short bursts of sound and using the same crystal to receive the echoes. This allows for the examination of deep structures using time-of-flight information. To avoid range ambiguities, we must analyze the return from one pulse before sending out the next. Aliasing problems may arise if the pulse repetition rate is too low. If the sampling is too low incorrect Doppler shift and directions are obtained. Following the Nyquist sampling theorem aliasing can be avoided.

8.4.3 Duplex ultrasound

Duplex systems use Doppler information superimposed on 'B'-mode diagnostic ultrasound scans. One way to accomplish this is by time-sharing. A fraction of the time is used to acquire a 'B'-mode scan and the other fraction is used to acquire the Doppler information. The crystal's resonant frequency for Doppler ultrasound is less than the resonant frequency for the 'B'-mode ultrasound. This may imply that different crystals should be used in duplex systems, but there are systems that use only one crystal successfully.

Linear arrays are very effective in duplex systems. An array of crystals can be composed of hundreds of elements (crystals). This allows the use of some elements for 'B'-scanning and others for Doppler scanning. The screen update for this technique is fast enough to fool the user into believing that both Doppler and 'B'-scan data are being acquired simultaneously.

8.4.4 Color flow Doppler ultrasound

Color flow Doppler (CFD) is not that different from the duplex system. Duplex systems show anatomical as well as flow information. Sometimes it is important to define the direction of the flow. Using phase V quadrature techniques the flow direction can be estimated. Depending on the direction of the flow a color can be assigned arbitrarily, e.g. red can be flow traveling towards the scan probe and blue away from it.

8.5 THREE-DIMENSIONAL (3D) ULTRASOUND

Three-dimensional ultrasound, better known as 3DUS, is receiving much attention. This is due to the benefits that 3DUS can provide, such as improved visualization, volume measurement, reduced examination time and analyses of the volume of interest (Shapiro *et al* 1999).

Figure 8.21 shows the block diagram of a 3D ultrasound system; this is just one of the many different systems available. To create 3D ultrasound images we use a regular ultrasound machine capable of generating 2D images and a sensor on the ultrasonic probe to provide information about the precise attitude of every scan. This information is loaded into a computer that processes it and generates the 3D image. Some of the position sensors used are articulated arms, spark gap microphones, laser and infrared sensors, and electromagnetic positioning systems (Nelson and Pretorius 1997).

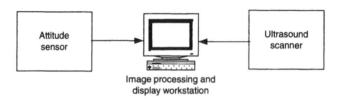

Figure 8.21. 3D ultrasound uses a 2D ultrasound scanner plus an attitude sensor.

8.6 BIOEFFECTS

Ultrasound is regarded as one of the safest imaging modalities available today. Ultrasound delivers energy in the form of sound waves into living tissue. Whenever energy is delivered into living tissue there is potential for some biological effects.

Many things have been said about negative effects from ultrasound exposure. On living human cells (leukocytes) single-strand breaks in the DNA after exposure to ultrasound have been observed, but the results have not been confirmed (Kremkau 1995). Sister chromatid exchange has also been studied. This refers to the exchange of portions of chromosomes in chromosome pairs. However, sister chromatid exchange does not introduce risk because it has no genetic effect.

Many studies have been done *in vitro*. According to the American Institute of Ultrasound in Medicine (AIUM), *in vitro* studies are very important to the understanding of the biological effects. However, reports from *in vitro* experiments claiming clinical significance should be viewed with caution but never dismissed. From animal studies negative effects such as fetal weight reduction, postpartum fetal deaths, fetal abnormalities, tissue lesion and many others have been reported (Kremkau 1995). Nevertheless, careful study should be made before extrapolating those results to humans.

The mechanisms used by ultrasound to create biological effects are heat and cavitation. An increase in temperature of less than 1 °C (from the normal temperature) is not harmful. In contrast an increase in temperature of more than 2–2.5 °C has been found to cause abortion and fetal abnormalities in multiple species. Sources that create heat are small transducers, high frequencies and focused beams, just to mention a few. Intensities of just a few hundred milliwatts per square centimeter can cause permanent damage. These toxic effects are a function of the duration of exposure, the proliferation rate, potential regeneration, intensity and others. It is also interesting to note that for temperatures of less than 2 °C, exposure for up to 50 h has shown no negative effects. In general no negative effects have been found for a given temperature T if the time of exposure is less than or equal to

$$t \leq 10^{6(6-T)} \tag{8.11}$$

given in minutes ranging from 1 to 250.

As mentioned before, the beam intensity is an important factor for the heating mechanism. The intensity has been limited by the Food and Drug Administration to 720 mW cm^{-2}: this intensity is computed in terms of spatial peak temporal average.

Cavitation refers to the excitation of gas bubbles present in human tissue. If the frequency is correct, these bubbles can resonate and grow. Sometimes shear stress and microstreaming may occur. Sometimes the expansion is so violent that the bubbles may collapse. However, little evidence of problems caused by cavitation has been found in humans.

In conclusion, there is no evidence that points to negative effects caused by the current diagnostic ultrasound. 'However, a prudent and conservative approach to ultrasound safety is to assume that there may be unidentified risk that should be minimized in medically indicated ultrasound studies by minimizing exposure time and output' (Kremkau 1995). For more information on ultrasound safety consult the American Institute of Ultrasound in Medicine homepage (www.aium.org).

PROBLEMS

8.1 Assume water and muscle have the same density. The velocity of sound in water is 1496 m s^{-1} and in muscle is 1568 m s^{-1}. Calculate the fraction of energy reflected.

8.2   Explain why the Q is different in a continuous wave and a pulsed ultrasonic transducer.

8.3   Calculate the distance of the near field for a 10 mm diameter, 2 MHz ultrasonic transducer.

8.4   Calculate the maximal time difference for a 10 mm wide phased array transducer to achieve a 30° angle.

8.5   Describe the factors that affect axial and lateral resolution for an ultrasonic transducer.

8.6   List the major sources of artifact for ultrasonic transducers.

8.7   Distinguish between pulsed Doppler, duplex Doppler and color Doppler ultrasonic measurement.

8.8   Calculate the maximal exposure time for a temperature rise of 5 °C.

REFERENCES

American Institute of Ultrasound in Medicine *Official Statements and Reports Concerning Safety and Bioeffects* webpage www.aium.org/stmts.htm

Kremkau F W 1995 *Doppler Ultrasound: Principles and Instruments* 2nd edn (Philadelphia, PA: W B Saunders)

Nelson T R and Pretorius D H 1997 Interactive acquisition, analysis and visualization of sonographic volume data *Int. J. Imaging Syst. Technol.* **8** 26–37 webpage tanya.ucsd.edu/papers/ijst/ijst_ppr.html

Quistgaard J U 1997 Signal acquisition and processing in medical diagnostic ultrasound *IEEE Signal Process.* **14** (1) 67–74

Shapiro I, Drapkin E, Halmann H, Shochet O and Graif M 1999 *Free-Hand 3D Ultrasound System* webpage http://www.cs.uwa.edu.au/~bernard/us3d/publications.html

Williamson M R 1996 *Essentials of Ultrasound* (Philadelphia, PA: W B Saunders)

CHAPTER 9

MULTIMODAL IMAGING

Alberto Rodriguez-Rivera

Multimodal imaging is an imaging method where multiple distinct anatomical imaging modalities as well as functional information are fused into a single framework. This process could be as simple as data overlapping or as complex as a nonlinear transformation of the data sets. Image registration, image segmentation, 3D-image generation, rendering and many more signal and image processing techniques are used to create these images. This image modality increases the overall understanding of the normal or abnormal behavior of tissue activity and is an excellent tool for the diagnosis and monitoring of patients.

9.1 MULTIMODAL IMAGING VERSUS IMAGE FUSION

The definition of image fusion is a subject of controversy. Fusion is defined as a merging of diverse, distinct, or separate elements into a unified whole. Therefore, image fusion should be defined as the merging of diverse, distinct or separate sets of modalities into a unified whole. This definition recognizes no difference between image fusion and multimodal imaging. However, others define image fusion as the union of multiple images to provide better pictures. For the sake of discussion we are going to deal with multimodal imaging (MMI) and image fusion (IF) as different applications.

9.2 MULTIMODAL IMAGING

Multimodal implies the merging of multiple modalities into a unified whole. One application that can explain this definition is multimodal neuroimaging.

Multimodal neuroimages convey information by merging anatomical as well as functional data. Anatomical data are extracted from MRI, PET or CT images while functional data are extracted from EEG and/or MEG studies. All this information is transformed and analyzed to create a new image. Figure 9.1 shows the block diagram of a multimodal neuroimager.

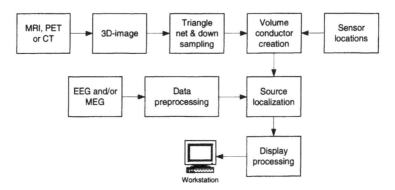

Figure 9.1. The anatomical data processing, first row, is performed only once per patient. Notice that there are two input blocks, one for anatomical pictures and the other for sensor information. The data preprocessing and source localization can be done multiple times depending on which areas are studied.

The applications for this image modality are countless. Two of the most important applications are the noninvasive 3D-source localization of epileptic foci for surgical removal and the assessment of brain functionality, e.g. after a stroke or trauma. The alternative to the noninvasive 3D-source localization of epileptic foci is the use of intracranial electrodes. Note that this procedure is mainly used for research purposes. Electrocorticography is the golden standard for source localization. Nevertheless, this methodology supplements electrocorticography and anything that helps this process is welcomed. To perform electrocorticography it is necessary to expose the brain in order to lay an array of electrodes on top of the cerebral cortex. On the other hand, brain functionality assessment encompasses many applications such as the 3D localization of pain centers, and the assessment of brain functionality, e.g. after a stroke, and many others. Multimodal imaging is not the only modality used for the assessment of brain functionality, as functional magnetic resonance imaging is available.

The process of fusing images from different modalities implies the use of a unified coordinate system. All discussions throughout the rest of this chapter assume that we are working on this new coordinate system. The method that transforms the old coordinate systems into the new one is called *image registration*. Image registration is widely used in areas such as virtual reality for the creation of 3D images from 2D images. Image registration is explained in more detail in section 12.1.1.

After registration, the first step towards the creation of this unified image is to process the anatomical data. The anatomical data (e.g. MRI scans) are transformed into a mathematical model that provides accurate geometrical information. Size, shape and width of the skull, shape and size of the cortex, size of the cerebrospinal fluid (CSF) compartment and much more information are gathered from this 3D model. This geometrical model together with other data is used to generate a *volume conductor model* (VCM). The VCM defines the linear relationship between energy sources in the cortex and the net magnetism or voltage at the sensor locations, which implies knowledge about conductance of the involved tissues. If the exact locations of the sensors are known, e.g. taken from a digitizer, they should also be included in the mathematical model to increase localization accuracy. This process is often regarded as a set-up step, because it is done only once for every patient.

The second step is to work with the functional data. Functional data are first preprocessed to extract artifacts. Some signal processing might be used to extract information from the signals that will help later in the process. Using this information, i.e. the functional data, the geometrical model from step one, the sensor locations and the VCM (also computed from step one), we can compute the location of cerebral activity inside the brain. This can be performed using a technique that minimizes the error of an index or with classical beamforming techniques. Let us study these steps in more detail.

9.2.1 Anatomical data and the volume conductor model

The MR scans provide the anatomical base for the VCM, but some transformation is necessary before they are useful. The first step is to create a real 3D version of the MR slices. This is accomplished by stacking the MR slices on top of each other. With knowledge of the size of every voxel, the width of every slice and the size of the space between slices we can construct a 3D image that conforms to the real dimensions of the patient.

After the 3D model is generated, we need to extract the shape of several anatomical structures from it. This is accomplished with a process called *segmentation*. Segmentation exploits the differences in the voxel's intensity levels generated by different anatomical structures, e.g. in an MR scan the skull is very bright while the CSF is very dark. By setting two intensity thresholds, one for high and one for low intensities, and selecting a seed (beginning) point inside the structure to be segmented, we can make an algorithm that travels from voxel to voxel comparing intensities until the thresholds are violated. After this process is done we have the envelope of a volume representative of the desired structure, as shown in figures 9.2(*a*) and (*b*).

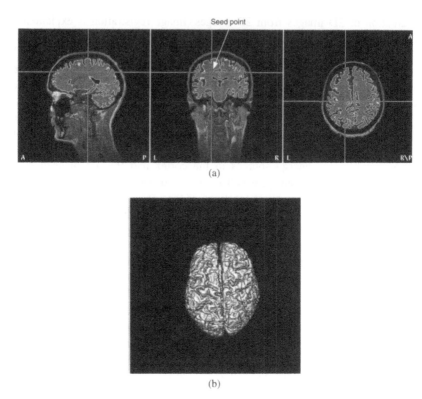

(a)

(b)

Figure 9.2. (*a*) Cerebral cortex segmentation performed on a 3D MR data set. The segmentation algorithm starts at the seed point and compares the intensity of the voxel with high and low voxel intensity thresholds. If the value is between the thresholds, the voxel is included as part of the segmented structure. The algorithm continues by selecting an adjacent voxel and repeating the process until it is violated. (*b*) Cortex segmentation results. The segmentation was performed on a 3D MRI image created from 124 raw MR slices (courtesy of the University of Chicago Comprehensive Epilepsy Center).

The VCM explains the linear relationship between an energy source inside the cortex and the potentials at the scalp electrodes assuming a particular head shape. As the energy travels outwards from its origin it comes into contact with tissues of different impedances, which attenuate the signal. There are several impedance zones that should be accounted for as a minimum requirement. They are the cortex–CSF, CSF–inner skull and outer skull–skin boundaries. All these zones are concentric and their boundaries are accounted for in the VCM; in other words, the VCM is a function of the head model (head shape). With a spherical head model approximation, i.e. concentric spheres, the VCM can be computed analytically for the EEG and MEG. To compute the VCM using a realistic head model, extracted from segmenting the MR scans, a numerical method should be used. The finite element method (FEM) and the boundary element method (BEM) are two examples of numerical methods to compute the VCM.

Spherical head models can be analytically computed, they are easy to implement and their execution time is rather fast. However, inaccuracies are introduced due to the significant difference between the spherical model and the patient's head shape. The realistic head model fits the patient's head shape almost to perfection. The cost of this gain in accuracy is a marked complexity increment and longer computation time. Figure 9.3(*a*) shows an example of a realistic head model and figure 9.3(*b*) shows an example of a one-sphere head model. These models are also called wire frame models.

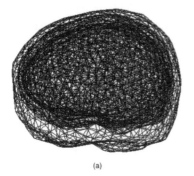 

(a)   (b)

Figure 9.3 (*a*) A realistic head model contains four concentric volumes representing the cortex, the CSF, and the inner and outer skull. When this model is overlaid onto the 3D MRI image an almost perfect fit is obtained. (*b*) A sphere representing the skin compartment is a very intuitive model. However, differences in shape create inaccuracies in the source localization. This model is often used for fast computations where accuracy is not a priority (courtesy of the University of Chicago Comprehensive Epilepsy Center).

After the shape has been extracted from the MR scans a triangle network (net) is used to represent this volume, just like the shapes in figure 9.3. The computation time and the memory to store the geometric representation are proportional to the number of triangle vertices. A large number of triangle vertices give good geometrical resolution but computation time increases as well. Down sampling or thinning is used to decrease the number of triangles and therefore save computation time and storage space.

9.2.2 Source modeling

Source modeling is very important for the success of source localization because the algorithm performance is as good as its model. This is also true for the head model. The source model is the mathematical expression that explains the electric or magnetic nature of brain cells. The so-called *equivalent current dipole* is one of the classic source models. It comprises a short, thin current path and the return currents in the surrounding material, i.e. short in terms of the distances between the sources and the electrodes.

Let θ be a *location* in a 3D space; you can see it as a vector in the space. We define $m(\theta)$ to be the 3×1 *dipole moment* at location θ, measured in µA mm^{-1}. Assume that we have an *M*-sensor EEG or MEG system. The potentials at the sensors caused by the moment $m(\theta)$ are given by

$$x(t) = H_\theta m_\theta(t) + n(t) \qquad (9.1)$$

where H_θ is the $M \times 3$ *lead field* matrix that relates the dipole moment at a given location and yields the potentials at the sensors. This is called the *forward problem*. Notice that this is a linear equation, hence, for multiple dipoles simultaneously active, the potentials at the electrode locations are given by

$$x = \sum_{i=1}^{L} H_{\theta_i} m_{\theta_i} + n(t). \qquad (9.2)$$

It is very important to understand that the lead field matrix H_θ contains information about the geometry of the head, the locations of the sensors and the conductances of every boundary. The computation of these matrices is not a simple matter; a good exposition is given by Gulrajani (1998). Equations (9.1) and (9.2) only explain the linear nature of the sources. These sources can be considered as random variables or they could be explained as chaotic processes. Loosely speaking chaotic means deterministic but complex and sensitive to the initial conditions.

9.2.3 Source localization

Source localization is also known as the *inverse problem*, i.e. given a set of time series (EEG or MEG waves at the scalp sensors) the dipole moment(s) that generated it must be found. This problem has been proven to have an infinite number of solutions. To solve it, constraints should be imposed to decrease the complexity, e.g. guess the number of dipoles and/or some behavioral characteristics that are discussed below. Hence, an optimization approach should be followed to find the closest feasible solution.

A classical approach is to minimize the *squared deviation* between the measured data x, i.e. the EEG, and the forward computed data Hm. The squared deviation, also known as the error, is given by

$$e^2 = \left\| x - \sum H(\theta_i) m(\theta_i) \right\|^2. \qquad (9.3)$$

Depending on the fixed variables, i.e. user-provided hints, we can come up with different dipole models. Most of them differ only by the computation of a different error measure.

9.2.3.1 Dipole models. Dipole models can be divided in two categories: *nondistributed* and *distributed*. Nondistributed dipole moments are the simplest to work with. The available variables are the location of the dipole(s), the moment and number of dipoles. From the available data (the EEG/MEG) the location of the dipole is fitted, and the moments are computed so that the *error* is minimized.

A *moving dipole* model is used when multiple data samples are selected and the location of the dipole(s) is not fixed. If the location of the moving dipole is fixed for a time range, the dipole can only rotate and the model is called a *rotating dipole*. If all the variables are fixed (held constant) except for the power of the dipole, we have a *fixed dipole* model. This concept can be extended to multiple dipoles. These are not the only choices for the model as *space scanning* is also available. With space scanning we compute or fit, for every dipole model, the best moment and error at preselected locations. This increases the computation time but yields confidence zones to make sense of the problem at hand. This method is computationally expensive, so it is necessary to predefine a subset of locations, such as a square grid, rather than computing at every possible location. Multisignal classification (MUSIC) is a scanning method that is very famous and widely used.

Distributed dipole models usually depend on current densities, measured in $\mu A\ mm^{-3}$. These current densities are discretized on a given set of locations, like a grid of points, by dipoles. This can be done because dipoles are proportional to the current density. The error measure for current density models differs from the other measures by an extra term called the *regularization* term. The regularization term is added to decrease the number of free variables and therefore obtain a unique solution. The error function becomes

$$e^2 = \|Hm - x\|^2 + \lambda\|Vm\|^2 \qquad (9.4)$$

where the regularization parameter, λ, controls which term dominates the error measure. If the parameter λ is large, the second term dominates the equation and large errors might be introduced. The idea is to balance it so the closeness achieved is not obscured by errors. The matrix V is used for depth bias control and for explicit weighting (Neuroscan 2000). Another interesting scanning method is low-resolution tomography (LORETA). The difference in the error estimation is in the regularization parameter, where the matrix V is replaced by the Laplacian. With this technique, neighboring dipoles possess similar strengths (bias control).

There are some issues concerning this methodology that should be mentioned. The user has the responsibility to select the number of dipoles that he or she believes effectively explain the data. The dipole model selection is also the responsibility of the user, i.e. moving, rotating or fixed dipole. If the user does not enter the correct information, large deviations from the correct result may occur.

It is difficult to understand the whole concept using only mathematics. Figure 9.4 explains the whole concept starting from the EEG. The lead field matrices H are computed before this process begins. We begin by acquiring an EEG and/or MEG. The user must provide some hints regarding the dipoles, such as the type of dipole and/or the number of dipoles present. From this information and the available data a dipole moment m is fitted. Then a fitted EEG is computed by multiplying the moment by the lead field matrix, Hm. The difference between this value and the real EEG is computed. Then the error is estimated and some parameters are recursively fitted. A new dipole moment is generated until the error is minimized.

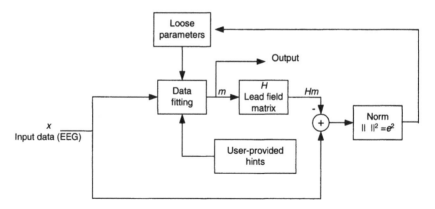

Figure 9.4. EEG data are entered into the system and one copy is stored for later comparison. The user must provide some hints that will increase the chances for a solution, e.g. number of dipoles and their nature, i.e. fixed, rotating, etc. A dipole m is fitted from the data and the hints. The error is computed and some parameters are changed to fit a new dipole moment. This process is repeated until the error is minimized.

9.2.3.2 Image resolution. Image resolution is not a problem for this algorithm. We can argue that the image spatial resolution is almost as good as the MR image resolution. However, this algorithm depends on a good SNR. A low SNR will result in large deviations from the correct answer.

9.2.4 Linearly constrained minimum variance (LCMV) spatial filters

LCMV is another signal-processing technique used for source localization. Spatial filters discriminate signals by their place of origin and not by their frequency content. This technique does not need user input about the dipole model or the number of dipoles. With good SNR levels the algorithm provides that information.

The formulation for this algorithm starts with the source model, equation (9.1) plus a white noise term. The dipole moment is a random quantity, therefore we can describe its behavior with its mean and variance. Let m_x and C_x be the mean and variance of x respectively. The spatial filtering process is linear, and is expressed as follows:

$$y = W^t(\theta_0)x \tag{9.5}$$

where W^t is the $3 \times M$ spatial filter for location θ_0, and M is the number of sensors.

The filter W should be designed such that

$$W^t(\theta_i)H(\theta_j) = \begin{cases} I, i = j \\ 0, i \neq j \end{cases} \tag{9.6}$$

i.e. signals from other locations should be annihilated while signals from the same location are passed without distortion. Realistically it is not possible to create filters with perfect transition bands and stop bands with no side lobes, and therefore the problem is expressed so that the filter's output variance is minimized. The problem is finally posed as follows

$$\min_{W(\theta)} \quad \mathrm{tr}(C_y) \quad \text{s.t.} \quad W^t(\theta)H(\theta) = I \tag{9.7}$$

where $\mathrm{tr}(A)$ means the trace of matrix A, i.e. the sum of the diagonal elements of A.

Equation (9.7) reads as to minimize the filter's output variance using the spatial filters as variables, maintaining the pass band intact if possible. The solution, obtained using the Langrange multipliers method, is given by

$$W_0^t = \left(H_0^t C^{-1}(x)H_0\right)^{-1} H_0^t C^{-1}(x). \tag{9.8}$$

Remember that $C(x)$ is a covariance matrix estimated with available data that are selected by the user.

9.2.4.1 Image generation. To generate an image, a dense constellation of spatial filters should be designed. Each filter provides power and direction information at its unique location. The variance at the filter's output is then computed and this serves as an index. This index is mapped into a color scale. The variance (index) at the filter's output is given by

$$\text{Var}(\theta_0) = \text{tr}\left\{ \left(H^t(\theta_0) C^{-1}(x) H(\theta_0) \right)^{-1} \right\} \tag{9.9}$$

where $C(x)$ is computed from available EEG data, and tr() is the trace of the resultant matrix. This is obtained by substituting equation (9.9) in tr($C(y)$). The variance can be computed as

$$C(x) = \frac{1}{(M-1)} \sum_{i=0}^{M-1} (x_i - m_x)(x_i - m_x)^t \tag{9.10}$$

where m_x, the statistical mean, can be computed as

$$m_x = \frac{1}{M} \sum_{i=0}^{M-1} x_i. \tag{9.11}$$

These results may be mapped into the 3D segmented surfaces for a realistic look, as shown in figure 9.5. If anatomical information is needed the results are thresholded and superimposed on a MR scan, very much like an fMRI. Often when the filters are sparse the output is interpolated and the brain is divided into axial slices where only surface maps are shown, as in figure 9.6. This technique is very simple but conveys much information. More elaborate data visualization techniques are being explored where virtual reality environments with stereoscopic imaging and real 3D capabilities are available.

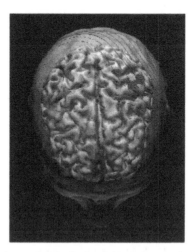

Figure 9.5. Source localization results may be mapped into segmented surfaces. The arrow shown in the left motor cortex represents the estimated dipole moment vector that explains the data or the minimization techniques or the filter with the largest variance. Notice that both algorithms can provide directional information of the dipole (courtesy of the University of Chicago Comprehensive Epilepsy Center).

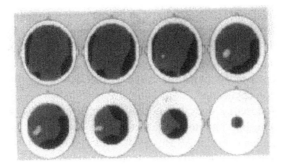

Figure 9.6. Source localization results may be presented as continuous energy planes. Bright zones represent high energy while black zones represent low energy (courtesy of the University of Chicago Comprehensive Epilepsy Center).

9.2.4.2 Resolution. Resolution is measured as the smallest distance between two sources that the system can differentiate. One source of problems in spatial resolution is coverage, which in turn affects the speed of computation as well as storage capacity. If you cover a volume with many filters, spatial resolution increases, but storage space used increases as well. Correlated sources are also known to create resolution problems, e.g. eye blinks. Mathematically speaking, correlated sources are related by

$$m_1 = am_2 + v \qquad (9.12)$$

where m_i is a dipole moment, a is a constant and v is a vector uncorrelated to the m_i. These types of source increase the number of missed detections due to averaging. On the other hand time resolution can be as fast as the sampling frequency, making this a good system for fast moving sources.

9.3 IMAGE FUSION

Image fusion is defined as the union of multiple images to create better pictures. Examples are MR and CT images (Hill *et al* 1991) and CT and SPECT images (Crum 1999). Notice that image fusion yields multimodal images that do not contain functional data input. Some image fusion techniques are image registration, image segmentation, 3D image creation from 2D slices, image and volume rendering and others. Notice that these techniques were presented for multimodal imaging. Therefore, we should focus on the description of some applications and results.

9.3.1 Virtual colonoscopy

Virtual colonoscopy is a good and virtually simple application to explain image fusion. Colonoscopy is the visual examination of the lining of the large intestine

(Chan 1999). The large intestine lining can be seen using an optical fiber camera inserted into the anus. An alternative that has been explored is called virtual colonoscopy. This procedure starts with an abdominal CT scan followed by some image processing performed on a general-purpose workstation. The fusion of multiple 2D images into a single 3D image is what qualifies virtual colonoscopy as an image fusion application.

Virtual colonoscopy does not necessarily rely on the creation of a new coordinate system, like multimodal neuroimaging, due to the fact that the data comes from the same modality. However, a new dimension is added to this coordinate system to host the 3D fused image. From the 2D slices, a 3D image should be created, just as explained with MR images in section 9.2.1. From this image we need to extract the colon. This is accomplished by image segmentation, also explained in section 12.1.1. The volume extracted is processed and finally rendered to create a virtual colon, as shown in figure 9.7. Virtual fly-through is performed to aid the visualization process. This is a very hot research topic and in the near future might be a *de facto* procedure for colonoscopies.

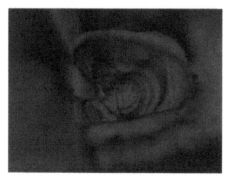

Figure 9.7. A virtual colon image created from a series of fused CT scan images (courtesy of the Colonoscopy Project of the Visualization Lab of the SUNY at Stony Brook).

9.3.2 Brain functionality with CT and SPECT

Section 9.2.1 describes multimodal imaging with an example of brain source localization. On that example brain functionality information was acquired by EEG and/or MEG data. Details such as source modeling and source localization algorithms should be kept in mind making the process a little complex. An alternative to this process is to use SPECT images to obtain functionality information.

Assume that we have CT and SPECT images available; see figures 9.8(*a*) and (*b*). CT and SPECT images provide redundant as well as complementary information. Fusing these two modalities is not simple. Differences in slice thickness, coordinate systems and resolution are just a few of the challenges to be solved. Figure 9.9 shows an example of the rendered volume resulting from the image fusion of three different data sets. Figure 9.10 shows another beautiful example of thoracic image fusion using 18-fluorodeoxyglucose (FDG) PET and CT scans.

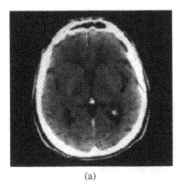 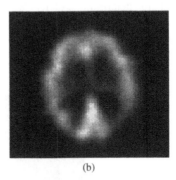

(a)   (b)

Figure 9.8. (*a*) CT brain image. (*b*) SPECT brain image. This shows how difficult it can be to diagnose a patient by looking at different modalities and to trying to identify the same location on both images.

Figure 9.9. A 3D head image with anatomical as well as functional data results from the fusion of three different data sets.

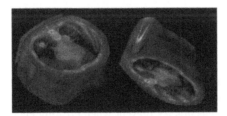

Figure 9.10. A 3D thoracic image from the fusion of 18-FDG PET and CT images.

9.3.3 Biomagnetic and bioelectric imaging

Bioelectric and biomagnetic imaging in their simplest forms overlay functional information on course approximations of biological structures, e.g. heart, thorax, head, etc. This process often does not convey anatomical information, making it difficult to determine exactly where the energy peaks are located. With anatomical scans, e.g. MRI or CT scans, the creation of a 3D model is possible. Data registration as well as image segmentation are used to create this model. Once the model is generated, the registered functional data can be easily overlaid on top of the segmented surface, as in the example shown in figure 9.11.

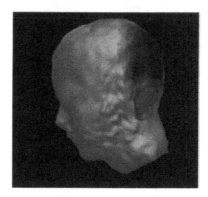

Figure 9.11. Bioelectrical image overlaid over a more realistic head shape. This image fusion aids in the visualization process by creating a color scale of power as a function of the head surface area (courtesy of Neuroscan Inc.).

9.4 BIOEFFECTS

There are no biological effects associated with multimodal imaging. They are passive data acquisition units, i.e. they only read biological signals without intervention. Nevertheless, the use of MRI, PET and CT has some bioeffects (see the chapters on MRI, PET and CT). Usually MRI, PET and CT scans are already available, which reduces any biological effects of this modality to zero.

PROBLEMS

9.1   List two examples of nonimaging sources used in multimodal imaging.
9.2   Describe 3D segmentation.
9.3   Describe the inaccuracies present when creating energy source models of the head.

9.4   Describe the method for determining source localization.

9.5   Explain the use of spatial filters.

9.6   Explain how virtual colonoscopy is achieved.

9.7   List the major problems in fusing CT and SPECT images.

REFERENCES

Chan I 1999 webpage www.gi-chan.com/colon.htm

Crum W R 1999 *3D Registration of SPECT and CT Brain Images* webpage
    http://agora.leeds.ac.uk/comir/research/brains/brains.html#comparison

Gulrajani R M 1998 The forward and inverse problems of electrocardiography *IEEE Eng. Med.*
    Biol. **17** (5) 84–101, 122

Haykin S and Steihardt A (ed) 1992 *Adaptive Radar Detection and Estimation* (New York: Wiley)

Hill D L G, Hawkes D J, Crossman J E and Gleeson M J *et al* 1991 Registration of MR and CT
    images for skull base surgery using point-like anatomical features *Br. J. Radiol.* **64** 1030–35

Maitre H and Block I 1997 Image fusion *Int. Sci. Found. Workshop Data Mining (Granada, Spain)*

Mosher J C, Lewis P S and Leahy R M 1992 Multiple dipole modeling and localization from spatio-
    temporal MEG data *IEEE Trans. Biomed. Eng.* **39** 541–57

Neuroscan Labs 2000 *Multi-Modal Neuroimaging User Guide*, Curry for Version 4.0, Sterling, VA,
    http://www.neuro.com/neuroscan/prod05.htm

VanVeen B D *et al* 1997 Localization of brain electrical activity via linearly constrained minimum
    variance spatial filtering *IEEE Trans. Biomed. Eng.* **44** 867–80

Wagner M and Fuchs M *et al* 1997 Automatic generation of BEM and FEM meshes from 3D MR
    data *Neuroimage* **5** S389

CHAPTER 10

GENERAL TECHNIQUES AND APPLICATIONS

Xu Li

The term *minimally invasive surgery* refers to surgical procedures and techniques that minimize the trauma of healthy tissue during operations. In many cases, the surgical site is accessed via small incisions and advanced surgical tools are employed to perform cutting, coagulation and vaporization with minimal injury to the surrounding tissue. The benefit of this type of surgery is obvious: it causes less pain, less blood loss and less stress to the patient; it requires less anaesthesia because of the smaller incisions; it avoids complications due to injury to healthy tissue; and it leads to shorter recovery time.

This chapter gives several examples of minimally invasive surgery to illustrate how the most advanced technologies are applied in this area. Chapter 11 provides a more detailed introduction to the scientific and engineering background.

10.1 MINIMALLY INVASIVE CARDIOVASCULAR SURGERY

Conventional open chest cardiovascular surgery involves cutting the sternum, separating the breastbones, stopping the heart and replacing the blood-pumping function by a heart–lung machine. This section introduces the techniques that enable the procedures for minimally invasive surgery. It reviews the latest advancements in treating coronary artery diseases, which make up a large portion of all cardiovascular conditions.

Coronary arteries are blood vessels that deliver oxygenated blood to the cardiac muscle. When these vessels become blocked or narrowed by fatty tissue or blood clots, the cardiac muscle is deprived of oxygen, which leads to heart attack or severe chest pain. Rather than performing an open chest operation, there are several alternative approaches to treating these diseases.

10.1.1 Minimally invasive direct coronary artery bypass

In order to restore the blood flow to the heart muscle, a procedure called *coronary artery bypass grafts* (CABGs) was developed. In this surgery, a healthy segment of blood vessel is attached in parallel with the blocked left anterior descending (LAD) artery or right coronary artery (RCA), providing an alternative pathway for blood flow to the cardiac muscle. Its less invasive counterpart is called MIDCAB (minimally invasive direct coronary artery bypass).

The vessel segment commonly used as a graft in MIDCAB is the left internal mammary artery (LIMA). Figure 10.1 shows that a specially designed endoscope (thoracoscope) is inserted through the fourth left intercostal space (ICS) and grafting instruments and scissors with electric energy are inserted through the third and fifth ICSs. The LIMA is harvested under the observation of the thoracoscope. Additional grafting instruments and scissors are inserted through third and fifth ICS to harvest the LIMA. Then, the grafting is performed under directed vitalization through a transverse incision at the level of the fourth ICS. The length of the incision is 5–10 cm. This procedure is performed under a beating condition and does not require the use of the heart–lung machine. Table 10.1 shows a comparison of MIDCAB to conventional CABG.

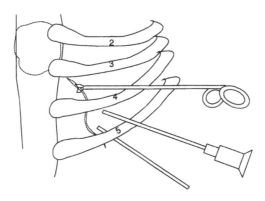

Figure 10.1. During a MIDCAB procedure, the harvesting of LIMA is under assistance of a thoracoscope, which is inserted through the fourth ICS. Additional grafting instruments and scissors are inserted through third and fifth ICSs to harvest the LIMA.

Table 10.1. Comparison of MIDCAB to conventional CABG surgery.

	Conventional CABG procedure	MIDCAB procedure
Incisions	Full sternotomy (removing of sternum bone) is performed. The average length of lesion is 30 cm.	The sternum bone is never fully broken. The incision is less than 10 cm long.
Heart condition	The heart is completely immobilized throughout the procedure.	The heart is beating during the procedure.
Heart function	Cardiopulmonary bypass pump (heart–lung machine) replaces the blood pumping function. Patients may suffer from complications due to the use of cardiopulmonary bypass pump.	The patient's heart and lungs continue to function independently. Heart–lung machine is not used.
Blood loss during the procedure	The patient has a higher chance of receiving a blood transfusion.	Generally blood transfusion is not needed.
Typical hospital stay	5–7 days.	2–4 days.

In order to perform a MIDCAB procedure, the following technologies and instruments are involved:

1. Local immobilization devices. Under a beating condition, the surgical site where the suture is performed must be locally immobilized. Reliable and accurate local immobilizers are developed to accomplish this function.
2. Thoracoscopes. Specially designed endoscopes that incorporate three-dimensional video display and computer motion control mechanism are used to visualize the surgical progress. Other surgical tools such as mini-occlusion clamps and mini-suturing devices are designed to perform the surgery through small incisions in the chest wall.

Because of its advantages, MIDCAB is becoming more and more popular in the United States. However, this new procedure is still in its evolutionary phase and various surgical approaches are being conducted on selected patient groups. A MIDCAB database and registry is being jointly established by the Society of Thoracic Surgeons and the American Association for Thoracic Surgery to facilitate the study of this technique.

10.1.2 PTMR

Coronary artery bypass is not the only means by which coronary artery diseases are treated. Studies show that small channels created in the inner wall of the cardiac muscle stimulate the formation of new blood vessels in the surrounding muscle. Oxygenated blood inside the ventricle perfuses into these newly formed vessels to alleviate the oxygen deprivation of the cardiac muscle.

PTMR (percutanous transmyocardial laser revascularization) is a minimally invasive surgical procedure based on such principles. During this procedure, the patient is under local anesthesia with a catheter inserted into the femoral artery at the top of the leg, and accessing the left ventricle through the aortic valve. Figure 10.2 shows that laser energy is transmitted through this catheter to create channels in the inner wall of the heart chamber.

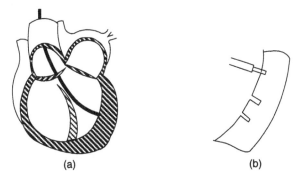

(a)   (b)

Figure 10.2. PTMR procedure. (*a*) The catheter goes through the aorta and is positioned within the left ventricle. (*b*) Laser energy is delivered to the tip of the catheter and cuts some small channels into the inner surface of the ventricular wall.

Figure 10.3 shows that laser energy is delivered by a laser catheter, which consists of an optical fiber, irrigation channel, electrodes and catheter hose. The diameter of the core of the fiber is about 0.4 mm.

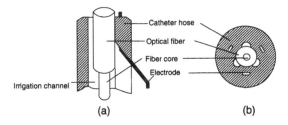

(a)   (b)

Figure 10.3. Laser catheter used in PTMR. (*a*) Longitudinal view. (*b*) Cross-sectional view.

The catheter is guided into the heart chamber under X-ray imaging and fluoroscopy. Advanced steerable catheters are designed to accurately position the tip of the catheter onto the inner wall of the heart chamber. The three electrodes also touch the inner ventricular wall and record the local electric potential continuously.

One of the lasers used in this procedure is a Nd:YAG laser providing continuous output power of 8 W at a wavelength of 1064 nm. Typically, 40–100 channels with a diameter of 0.5–1.2 mm and a depth of 4–11 mm are made into the inner ventricular wall.

10.1.3 Percutaneous transluminal coronary angioplasty

PTCA (percutaneous transluminal coronary angioplasty) is another minimally invasive catheter-based surgery to treat coronary artery diseases. In this type of procedure, the catheter accesses the coronary artery through the femoral artery, which is quite similar to the PTMR procedure. The difference is that a laser, balloon or other mechanism directly opens the blockage in the coronary artery.

Percutaneous transluminal balloon angioplasty has been performed with 2 cm long unidiameter balloons for almost 20 years. The uninflated balloon is guided by catheter to the narrowed coronary artery, then it is inflated and flattens the plaque on the inner arterial wall with pressure. The balloon has been made of compliant material (polyolefin copolymer), semicompliant material (polyethylene or polyvinyl chloride), or noncompliant material (polyethylene terephthalate). Only recently are compliant and noncompliant materials combined together to make balloons with a compliant center section and noncompliant ends. This design enables the pressure to be delivered focally and avoids damage to surrounding reference tissue.

In many modern balloon angioplasty procedures, a specially designed stent is used to improve the effect and eliminate complications. The stents are left in the coronary artery after angioplasty to prevent future occlusion. Figure 10.4 illustrates a stent-assisted balloon angioplasty procedure.

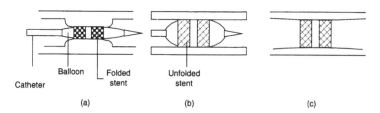

Figure 10.4. Balloon–stent angioplasty. (*a*) A catheter with a balloon on its top is inserted into the blood vessel. The balloon has a sharp tip installed on its top, making it easier to go though the occluded vessel. It is also equipped with a folded stent on it. (*b*) When the balloon inflates, the build-up within the blood vessel is flattened. The stent is unfolded by the balloon. (*c*) The balloon catheter is removed. The unfolded stent is left inside the vessel to support it and prevent it from further occlusion.

The stents for coronary arteries normally have diameters of 3.0–5.0 mm and lengths of 10–40 mm. There is a variety of different wiring styles, such as wire mesh, sinusoidal helical coil, multiple rings with multiple links and so on. Most currently available stents are made of stainless steel and the long-term effects of the implantation in the arteries are not clear yet. Stents coated with polymer, biocompatable material, immobilized drug, or radioactive material, which can selectively kill proliferating cells, are under experimental tests or have been suggested.

A blade-based angioplastic instrument has also been developed. This instrument has a rotational cutter installed on the catheter tip. It can eliminate the plaque by cutting or grinding. This technology shows the possibility of opening the occluded vessel more efficiently and more thoroughly.

10.2 MINIMALLY INVASIVE BRAIN SURGERY

The brain is an extraordinary complex organ made up of delicate and vital tissue. Minimally invasive brain surgery (MIBS) is one of the most attractive and challenging technologies for both physicians and biomedical engineers. The advances of MIBS include computer-assisted stereotaxis, intraoperative ultrasound, brain mapping and neuro-endoscopic microsurgery.

10.2.1 Endoscopic neurosurgery and endoscope-assisted microneurosurgery

10.2.1.1 Endoscopic pituitary tumor removal. Neurosurgery provides new applications of endoscopic technology. Endoscopic pituitary tumor removal has been a particularly successful example of endoscopic nerosurgery. The conventional surgical procedure for removal of pituitary tumors requires an incision under the upper lip in the nostril, a resection of the osseous septum and usually requires the use of postoperative nasal packing. The resection of the tumor is performed under microscopic magnification and fluoroscopic imaging. Because the view of the surgical site is not clear enough, sometimes an accident occurs with injuries to carotid arteries or optic nerves.

With the help of a small endoscope, this procedure can be done less invasively and more safely. During an endoscopic pituitary surgery, a 4 mm rigid endoscope is inserted into one of the patient's nostrils, passes through the sphenoid sinus, the anterior wall of the sella, the dura mater, and finally accesses the surgical site. Surgical instruments are placed in the same nostril to remove the tumor under the visualization of the pituitary tumor by the endoscope. No lip or nasal incisions are required in this procedure. The patients are generally sent home the day after surgery.

10.2.1.2 Endoscope-assisted microsurgery. Endoscope-assisted microsurgery for the brain is one of the latest advancement of MIBS. In a common microsurgical operation of the brain, a microscope is employed to magnify the image of the surgical site. However, microscopes suffer from a reduction of light intensity for deeper operating fields, and they provide rather narrow viewing angles. Thus, it is hard to observe objects located in the shadow of the microscope beam and normally further resection of bone edges and retraction of neuronal or vascular structures are required. The use of endoscopes helps to provide more visual information about the operating field, such as precise localization of the target, accurate topographical relations of specific lesions to individual anatomic variations of intracranial structures, and hidden corners of the operating field without causing further trauma to the brain tissues.

In endoscope-assisted microsurgery, the endoscope is held by a mechanical arm to the desired position. Rigid endoscopes with a diameter of 2–5 mm and fiber endoscopes with diameter of 1.0–1.5 mm have been used. The visualization of the surgical site is provided by the video camera attached to the endoscope and the image is displayed on a monitor.

Using the endoscope as an additional viewing instrument during microscope-based microsurgery means those two sets of visual information, microscopic and endoscopic, must be simultaneously integrated to gain complete information about the operating field. This task could be achieved by image fusion.

10.2.2 Image-guided stereotaxic brain surgery

Stereotaxis refers to a system of navigating to any point within the brain, with the aid of imaging techniques that concurrently display external reference landmarks or neural structures (including the brain lesions). The imaging data (with CT, MRI or PET) is integrated with the view of the surgical field with external head frames, or recently, with infrared optical probes, ultrasonic sensor assemblies or other mechanisms that do not require conventional stereotaxic frames. This provides reference points from which a computer can calculate and present trajectories and depths to any target point within the imaged brain. Detailed knowledge of the image-processing algorithm is given in chapter 11 (multimodal imaging) and chapter 15 (image guidance during surgery).

The steps below illustrate a procedure assisted by an ultrasound-controlled stereotaxic system to remove deep-seated tumors in the brain.

1. System set-up and calibration. The patient's head is placed in a head holder. The neuronavigator is held by a robotic arm with 6 degrees of freedom, which is controlled by a workstation. It is equipped with an ultrasound scanner with a 5 MHz sector or convex transducer, and other instruments such as needle pointers, endoscopes and biopsy forceps. The navigator is calibrated by pointing the needle pointer to a mark made on the head holder. The corresponding location on the CT or MRI scan is displayed by a cursor on the image.
2. Opening of burr hole. The needle pointer is moved along the scalp to the location of the tumor. The final position of the pointer is decided by the analysis of the CT/MRI data. Once the desired location is reached, a burr hole into the scalp is made at that location.
3. Biopsy. Incision into the brain tissue is made by biopsy forceps. CT/MRI data are used to decide how deep the biopsy forceps should go.
4. Ultrasound imaging. Any further incision should be confirmed with ultrasound images. The ultrasound images are made by sector and convex

probes, and are compared to the CT/MRI images to verify the location of the biopsy forceps, and also to provide more accurate and up-to-date information on the surgical site, such as vascular structures, newly made incision and bleeding.

5. Performing further biopsy or resection. Based on the ultrasound and CT/MRI images, further biopsy or resection is performed at the optimal location to remove the tumor.

Such a stereotaxic system enables the surgeon to reach a brain tumor or vascular malformation with minimal disruption of the overlaying brain.

10.3 MINIMALLY INVASIVE OPHTHALMIC SURGERY

The recent development of minimally invasive ophthalmic surgery is largely based on the advancement of computer and laser technology. Computer-aided laser ophthalmic surgery has distinct advantages over conventional ophthalmic surgery.

One reason relies on its precision and safety. Another reason is that the laser light can pass through the cornea to be concentrated inside the eye, therefore a surgical opening does not need to be made in the eyeball. These advantages have greatly decreased the risk of infection and reduced recovery time.

Chapter 15 presents the fundamentals of lasers. This section gives an introduction to two important applications of laser ophthalmic surgery—glaucoma surgery and laser cornea surgery.

10.3.1 Laser glaucoma surgery

Glaucoma refers to the increase of the intraocular pressure (IOP), which causes damage to the optic nerves. The vision may be impaired gradually and sometimes the condition progresses to blindness. This IOP results from an excessive amount of fluid pressure within the eye, which is caused by either excessive fluid production or deceased fluid drainage due to malfunction of the plumbing system. There are two types of glaucoma—open angle glaucoma and angle closure glaucoma. The former is the most common type and makes up more than 90% of all cases of glaucoma.

The purpose of surgical treatment of glaucoma is to make a new opening in the eyeball for the fluid to leave the eye. In conventional glaucoma surgery, the surgeon removes a small piece of tissue from the sclera (the white fibrous membrane that covers the eyeball), then covers the hole with conjunctiva, which is a transparent membrane covering the eyeball and under surface of the eyelid. The fluid flows through the new opening, under the conjunctiva, and drains from the eye. Because the globe is opened surgically, this operation might cause some side effects, such as cataract, inflammation or infection inside the globe, and swelling of blood vessels behind the globe.

Laser surgery for glaucoma is one of the most promising and successful of advances of glaucoma surgery. It largely reduces the risks described above. Today, most glaucoma patients can be effectively controlled by either medication or laser surgery.

10.3.1.1 Laser trabeculoplasty. During the past decade, a procedure called laser trabeculoplasty has been developed based on the work of Wise and Witter (1979). This procedure has greatly improved the treatment of open-angle glaucoma by performing the surgical procedure in a far less invasive manner.

In an argon laser trabeculoplasty (ALT) surgery, short bursts of laser energy are delivered to the trabecular meshwork, which is a spongelike ring of tissue located where the iris connects to the cornea. It is postulated that the laser burns cause shrinkage to the fibers of the spongy tissue of the angle, which then opens up drainage spaces, permitting the excessive fluid to flow through this area to be absorbed by the sinus veins.

For standard ALT surgery, laser burns are directed at equal intervals on the anterior portion of the trabecular meshwork. Normally the laser is in continuous-wave mode; the wavelength is set at the blue–green biochromatic wavelength spectrum ($\cong 50$ μm), with power of 0.6–1.2 mJ per pulse and the duration of the burns set to 0.1 s. Typically 50–100 burns with a diameter of 50 μm are made over 180–360° on the trabecular meshwork.

In order to focus the laser and visible light to the trabecular meshwork and help the surgeon obtain an enlarged view of the inside of the eye, a contact lens with mirrors, which is called a Ginio lens, is placed on the patient's globe as shown in figure 10.5.

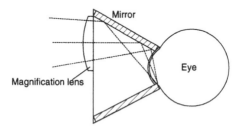

Figure 10.5. Ginio lens used in ALT surgery.

After the procedure, 70–80% of eyes respond favorably. The average reduction of pressure has been estimated as 7–10 mmHg.

10.3.1.2 Laser peripheral iridotomy. For angle-closure glaucoma, which is an unusual type of glaucoma, a laser is used to create holes in the iris, allowing the fluid to flow to the trabecular meshwork through a more direct pathway. This procedure is called laser peripheral iridotomy. Both the argon laser and Nd:YAG lasers perform this procedure well. Attention is now being focused on newer lasers, such as the semiconductor diode laser and the Nd:YLF laser.

10.3.2 Laser corneal reshaping surgery

The conventional surgical procedure that reshapes the cornea to correct vision is called radial keratotomy (RK), during which incisions are made in a radial pattern in the cornea using a hand-held surgical knife. This procedure is effective for changing the shape of the cornea. However, because the incisions are deep (sometimes 90% of the thickness of the cornea, see figure 10.6(*a*)), and the healing process varies greatly among individuals, there are considerable chances for complications. For example, the weakening of the cornea frequently leads to progressive flattening of the cornea and hyperopia develops. More serious complications include severe scarring and constant blurring of vision.

(a)   (b)

Figure 10.6. The cornea correction profile performed by (*a*) RK and by (*b*) PRK.

With the development of laser technologies, RK is rapidly being replaced by laser surgery for almost all vision correction applications. Laser surgery is much safer and more precise than RK, and eliminates the serious complications of RK. Today, laser photorefractive keratectomy (PRK) is a standard procedure performed in clinics to treat myopia, astigmatism and hyperopia.

10.3.2.1 Photorefractive keratectomy. PRK uses a laser to reshape the cornea in order to change its ability to focus light on the retina. Since 1995, several laser systems have been approved by Food and Drug Administration to treat myopia and astigmatism.

Instead of making radical cuts in the cornea, figure 10.6(*b*) shows that the PRK procedure employs the laser beam to reshape the cornea by sculpting an area with diameter of 5–9 mm in the center of cornea. This process removes only 5–10% of the thickness of the cornea for mild to moderate myopia and up to 30% for extreme myopia so that the integrity and the strength of the corneal dome are retained.

The laser used in PRK is an excimer laser with a wavelength of 193 nm because its short pulse duration (12–15 ns) confines the thermal effects to infinitesimal levels. Thus the incisions can be made accurately with little trauma to the surrounding tissue. The advanced beam delivery system allows the intensity distribution to be delivered over an area with a diameter of 7.0 nm. Figure 10.7 shows that an aperture wheel controlled by computer is employed to adjust the beam diameter, which determines the contour of the sculpted area.

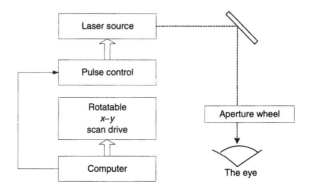

Figure 10.7. The system diagram for the PRK procedure.

Figure 10.8 shows that the myopia aperture wheel has a series of apertures of decreasing diameter. More laser energy is applied to the central cornea compared to the peripheral cornea. The speed of the wheel is controlled by computer, which determines the degree of correction. The aperture wheels for hyperopia and astigmatism have more complicated apertures.

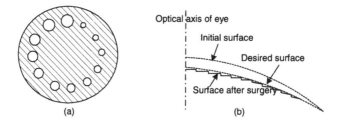

Figure 10.8. The myopia aperture wheel (*a*) and the myopia correction profile (*b*).

FDA research showed that for mild to moderate myopia, 95% of the patients have vision of 20/40 or better after surgery, and approximately 2/3 of the patients had 20/20 vision. Risks of PRK are minimal. In a small percentage of cases, the patient experiences a slight corneal 'haze' or has symptoms of glare in bright light.

10.3.2.2 Laser in situ keratomileus. An even less invasive procedure than PRK is called laser *in situ* keratomileus (LASIK). In a PRK procedure, the Bowman membrane, which refers to the protective layer on the outer part of the cornea, is removed, whereas in LASIK, this layer is preserved. During this procedure, a flap of the Bowman's layer is cut with a knife called a microkeratome. The targeted tissue beneath it is removed with the laser (in a similar manner as for PRK), and then the Bowman's layer is repositioned.

Because the Bowman's layer is preserved in LASIK, postoperative pain is substantially reduced and a shorter recovery time to restore vision is required. In the long term, there is less chance of corneal scarring or change due to healing, thus bringing greater stability of the correction. However, due to the complexity of the procedure, LASIK has yet to be approved by the FDA.

PROBLEMS

10.1 List the advantages of minimally invasive surgery. List the challenges it brings to surgeons and biomedical engineers.

10.2 Explain the difference between PTMR and PTCA.

10.3 The aperture wheel in figure 10.9 is used in a PRK procedure. Select whether it is for treatment of myopia, hyperopia or astigmatism.

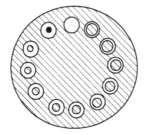

Figure 10.9. An aperture wheel used in PRK surgery.

REFERENCES

Benson W E, Coscas G and Katz L J (ed) 1994 *Current Techniques in Ophthalmic Laser Surgery* 2nd edn. (Philadelphia, PA: Current Medicine)

Cohen R G, Mack M J, Fonger J D and Landreneau R J (ed) 1999 *Minimally Invasive Cardiac Surgery* (St Louis, MO: Quality Medical)

Danial J and Dattolo J 1998 Minimally invasive cardiac surgery: surgical techniques and nursing considerations. *Crit. Care Nursing Quart.* **20** (4) 29–39

Diegeler A, Falk V, Krahling K, Matin M, Walther T, Autschbach R, Battelini R and Mohr F W 1998 Less-invasive coronary artery bypass grafting: different techniques and approaches. *Eur. J. Cardio-Thoracic Surg.* **14** (Suppl 1) S13–19

Fisher P E, Khomoto T, DeRosa C M, Spotnitz H M, Smith C R and Burkhoff D 1997 Histologic analysis of transmyocardial channels: comparison of CO_2 and holmium:YAG lasers *Ann. Thoracic Surg.* **64** (2) 466–72

Hodgson J M 1997 Focal angioplasty: theory and clinical application *Catheterization Cardiovasc. Diag.* **42** (4) 445–51

Jho H D, Carrau R L, Ko Y and Daly M A 1997 Endoscopic pituitary surgery: an early experience *Surg. Neurol.* **47** (3) 213–22

Karlin D B (ed) 1995 *Lasers in Ophthalmic Surgery* (Cambridge, MA: Blackwell Science)

Koivukangas J, Louhisalmi Y, Alakuijala J and Oikarinen J 1993 Ultrasound-controlled neuronavigator-guided brain surgery *J. Neurosurg.* **79** (1) 36–42

Perneczky A and Fries G 1998 Endoscope-assisted brain surgery: part 1—evolution, basic concept, and current technique *Neurosurg.* **42** (2) 219–24

Ruzzu A, Fahrenberg J, Muller M, Rembe C and Wallrabe U 1998 A cutter with rotational-speed dependent diameter for interventional catheter systems *Proc. MEMS 98 IEEE, 11th Annual Int. Workshop Micro Electro Mechanical Syst.* (New York: IEEE) pp 499–503

Scarlett M V 1998 Minimally invasive cardiac surgery: a new frontier *Crit. Care Nursing Quart.* **21** (1) 16–23

Takahashi M, Yamamoto S and Tabata S 1997 Immobilized instrument for minimally invasive direct coronary artery bypass: MIDCAB doughnut *J. Thoracic Cardiovasc. Surg.* **114** (4) 680–82

Violaris A G, Ozaki Y and Serruys P W 1997 Endovascular stents: a 'break through technology', future challenges *Int. J. Cardiac Imaging* **13** (1) 3–13

Weber H P, Heinze A, Richter U, Ruprecht L, Zhuang S and Unsold E 1998 Transcatheter endomyocardial laser revascularization: a feasibility test. *Thoracic Cardiovasc. Surg.* **46** (2) 74–6

Wise J B and Witter S L 1979 Argon laser therapy for open angle glaucoma, a pilot study. *Arch. Ophthalmol.* **97** 319

Zamorano L J, Nolte L, Kadi A M and Jiang Z 1993 Interactive intraoperative localization using an infrared-based system *Neurol. Res.* **15** (5) 290–98

CHAPTER 11

ENDOSCOPIC SURGERY

Xu Li

During many minimally invasive surgical procedures, surgical tools are inserted through small incisions or even natural openings of the body. In order to observe the manipulated tissue and the surgical progress of the surgery, an endoscope is also inserted into the surgical site. Endoscopes are designed in different sizes and styles according to applications. Some of the various types include:

1. *Laparoscope*. Designed for viewing in the abdominal area; applications include gall-bladder operations, hernia repair, appendix removal, etc.
2. *Arthroscope*. Designed for viewing inside joints; applications include knee surgery, shoulder surgery, wrist surgery, or even spine surgery.
3. *Sinuscope*. Designed for viewing nasal and sinus cavities; used in sinus surgery.
4. *Cystoscope*. Designed for viewing the bladder; used in prostate surgery and other procedures related to the urinary and reproductive systems.
5. *Hysteroscope*. Designed for viewing the uterus and assisting with certain gynecological procedures.
6. *Colonoscope*. Used for viewing the colon.

Many modern endoscopic systems are equipped with video cameras and monitors, which are used to display the magnified and processed video images, thus providing visual information more effectively to the surgeons and their teams. Also the endoscope might be incorporated with an operating mechanism, such as a laser or electrosurgical knife, to perform surgery at the observed site directly.

This chapter introduces endoscopes from an engineering point of view. In particular, two important examples of endoscopic surgery—laparoscopy and arthroscopy—are described in detail to illustrate the applications of advanced technologies in endoscopic surgery.

11.1 ENDOSCOPES

Endoscopes provide visualization of the manipulated tissue and the progress of the procedure during any type of endoscopic surgery. Although the sizes and styles of endoscopes vary from case to case according to the application, each type shares the same principles and overall structure. There are two major categories of endoscope: rigid and flexible scopes.

11.1.1 Rigid endoscope

Typically, a rigid endoscope consists of a tube that contains the optical system, object lenses at the distal end, eye lenses at the proximal end, and an optical relay system located between them. The dimension of a rigid endoscope ranges from 0.1 cm to 2 cm in diameter and from 10 cm to 1 m in length.

The optical relay system passes the image from the distal end to the proximal end and also corrects the aberration produced by the object lens. Most modern rigid endoscopes employ Hopkins rod lenses as the relay system. As shown in figure 11.1(*b*), a Hopkins lens system is composed of quartz or plastic rods with spherical and flat ends, which create an *air lens* in the middle. Compared to the conventional relay system consisting of thin biconvex lenses (figure 11.1(*a*)), the Hopkins lens system has fewer reflecting surfaces, thus increasing light throughput and reducing vignetting (darkening of the off-axis portion of the image). Also the rod-shaped lenses enable reduction of the diameter of the endoscope and decrease the chances of dislodging a lens during handling. A new type of relay system—gradient index (GRIN) rod lens system—has even fewer reflecting surfaces. The light path bending within the rod lens is caused by a radial variation of optical index of the material. The manufacturing process of this type of rod lens involves diffusing dopants radially into the rod to establish an optical index gradient.

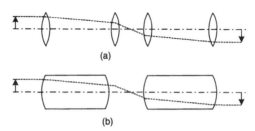

(a)

(b)

Figure 11.1. (*a*) Conventional biconvex lens relay system. (*b*) Hopkins lens relay system. The Hopkins optical system is composed of rod lenses.

The illumination for a modern endoscope is provided by cold light sourced from halogen, xenon or halide lamps. The light is transmitted to the proximal end by a few thousand glass fibers located on the periphery, each about 70 μm in diameter.

Figure 11.2 shows the complete optical system for a typical rigid endoscope. The object lens system is composed of a negative lens and two biconvex lenses, which magnifies the view of the object depending on the distance from the object. The Hopkins relay system consists of two or more pairs of rod lenses, depending on the size of the scope. The eyepiece contains two convex lenses and is often coupled with a color CCD (charge-coupled device) camera. The observed view is displayed on one or more monitors to give visualization of the surgical view to the surgeon and his/her team.

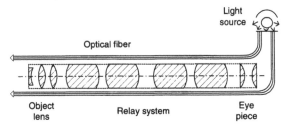

Figure 11.2. The eye on the right looks through the optical system in a rigid endoscope. The light source transmits light through the surrounding optical fibers to the object at the left.

To increase the viewing area of an endoscope, sometimes the endoscope is designed with a viewing angle by adding a prism, a mirror, or a curved light guide at its distal end. Figure 11.3(*a*) shows that for such an endoscope, the overall viewing area can be increased by rotating it. Figure 11.3(*b*) shows that an endoscope with a wider angle allows a broader view but also has greater distortion.

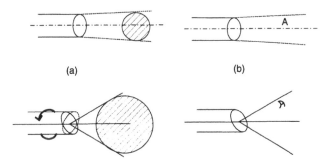

Figure 11.3. (*a*) The viewing area of an angled endoscope and (*b*) the observed distortion.

11.1.2 Flexible telescope

The advantage of a rigid endoscope is that it has a much simpler structure than a flexible endoscope and provides satisfactory imaging quality more easily, due to the aberration correction function of the relay system. The drawback is that a rigid scope cannot bend around corners to access a hidden area because of its rigid structure. The flexible endoscope, on the other hand, can be inserted along a complicated path, and enable the surgeon to view the object in any direction. The bronchoscope, gastroscope and colonoscope are the most important flexible endoscopes.

The relay system of a flexible endoscope is a fiber bundle. The illuminating light is also carried by fiber bundles located in the same tubular-shaped casing but separated from the image relaying bundles. Typically, the diameter of individual fibers is 10–100 μm. The light fibers usually form a semicircular cross-section configuration and a hollow duct is also included in the casing to provide a pathway for by-products of the surgery such as fluid and gases. Figure 11.4 shows the intersection of the relaying and light carrying fiber bundles.

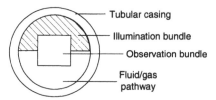

Figure 11.4. The relay system for a flexible endoscope contains a pathway for injecting and removing fluids.

An optical fiber bundle lacks the aberration correction ability of the Hopkins rod lens system. Instead, because of the light throughput loss and honeycomb patterns formed by individual fibers within the bundle, the resulting image quality is degraded. As a result, it often requires a complicated postprocessing system to obtain clear, true color images. The digital images picked up by the CCD camera are processed by specially designed algorithms before they are displayed on the monitor. Figure 11.5 shows a flexible endoscope system.

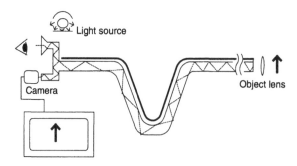

Figure 11.5. A flexible endoscope can follow a complicated path.

In a new type of flexible endoscope configuration, minimized CCD cameras are placed at the distal end of the scope and the digitized image is passed through signal wires to the display terminal (figure 11.6). In one particular design, only a single monochrome chip is used instead of using three chips and separate color filters. The color image is formed by alternating the color of the illumination. The system using a CCD to obtain an image at the distal end has high resistance to noise, virtually no transmission loss and less image distortion.

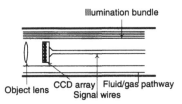

Figure 11.6. A flexible endoscope system with CCD camera requires a lens for imaging.

11.1.3 New developments and perspectives of endoscopic technology

In addition to the conventional optical approach, other modalities of imaging technology, such as ultrasound and MRI, have been or are going to be incorporated into endoscopes. In order to obtain microstructural information, various technologies of microscopy have been employed in endoscopes.

Laser-based optical technology makes it possible to obtain the image of individual cells inside the body. For example, endoscopes using laser-scanning confocal microscopy technology scan the tissue pixel by pixel and may achieve a resolution of 0.5–5 µm. This will give histological information on the surface cells at the distal end of the endoscope.

Another emerging technology is called 'optical coherence tomography' (OCT). In OCT-incorporated endoscopy, laser beams are delivered by optical fibers to the tissue and back-scattered light is collected by a procedure called 'low-coherence interferometry'. Then the back-scattered light is used to reconstruct the image of the tissue in a way similar to B-mode ultrasonic imaging. OCT can achieve a high resolution (0.1 μm) and a penetration depth of 2–3 mm (cell layers). Thus a subsurface histological image could be obtained for pathological analysis of the tissue.

Spectroscopy may also be employed in endoscopy, in which illuminating light is delivered to the tissue being examined and the spectra of the light scattered or emitted by tissue is analyzed to identify tissue composition and pathology. The excitation light is delivered by only a few optical fibers and the scattered or emitted light can be collected by optical means or by miniaturized CCD cameras. The whole system will be compact and when it is integrated with flexible endoscopy, it is possible to examine the cervix or colon for premalignant conditions without surgical biopsy.

Another improvement of endoscopic technology is to present stereoscopic visual information to give the surgeon a perception of depth, which significantly facilitates complicated procedures such as suturing or knotting. The basic idea of the stereoscopic endoscope is to take images from left and right viewpoints, and then to display them sequentially on a monitor. Wearing active multiplexing glasses, the surgeon receives a 3D sense of view.

A standard stereoscopic endoscope employs two lenses and two cameras to gather the left and right views. Figure 11.7 shows that in order to reduce the size of the endoscope, a system with only one lens and one CCD camera was developed. Figure 11.7 shows that this system switches between the left and right views by controlling the optical axis direction of a ferroelectric crystal device, which is located between the lens and the CCD camera.

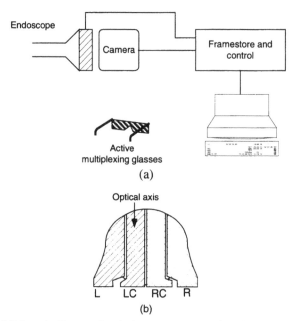

Figure 11.7. (*a*) Schematic diagram of a single-camera stereoscopic endoscope. The left view and right view are switched by controlling the optical axis direction of a ferroelectric crystal device. (*b*) A ferroelectric crystal device controlled by four electrodes. The optical axis direction of the crystal and the proportion of lens that is used to form the image on display can be selected electrically. L-LC driven: left view is selected. R-RC driven: right view is selected.

11.2 MECHANICAL SURGICAL TOOLS FOR ENDOSCOPIC SURGERY

In endoscopic surgery, the internal organs are manipulated by tools inserted through small incisions or natural openings of the body. Because of the restraint on both access and visualization, the surgical tools for endoscopic surgery are specially designed to be smaller, more precise, and with more control. This section discusses the mechanical issues of the surgical tools.

11.2.1 Endoscopic surgical tools for dissection, ligation and suturing

Most mechanical surgical tools for endoscopic surgery consist of finger loops or handles outside the patient's body, a tool tip inside the patient's body, and a long tube or shaft with an internal mechanism relaying force and control from the handle to the tip. According to the application, the length of the shaft ranges from 15 cm to 50 cm. For almost every category of surgical tools, both reusable and disposable versions are available. Disposable instruments are gaining more publicity because their sterility is guaranteed.

Endoscopic scissors are used for making incisions in different tissues, cutting blood vessels, ducts, or suture material. Currently, 5 mm endoscopic scissors are available in different lengths. They can rotate 360° and flex for 80° to dissect delicate structures in areas difficult to access.

More complicated operations such as knotting and suturing can also be performed in ways similar to those in open surgery. Obviously more expertise and operation time are required for these operations. In many situations, alternative approaches can be used. For example, ligature clips can be used to close vessels or ducts and fix terminal ends of sutures. The clips are made of stainless steel, titanium or absorbable plastic. They are available in different lengths ranging from 5 mm to 11 mm. The clip appliers can fire 15 to 30 clips in rapid succession and automatically load new clips. Stapling devices can be used to replace sutures. Figure 11.8 shows a stapling device that cuts and secures the tissue in a single operation.

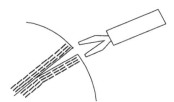

Figure 11.8. An endoscopic stapling device places six rows of staples, then cuts between the middle rows.

11.2.2 Haptic feedback for endoscopic surgery

Haptic perception refers to the combination of tactile and kinesthetic information. During endoscopic surgery, the surgeon's perception of the cutting progress is limited to visual feedback of the laparoscope. Haptic perception, which is very important to a trained surgeon, is lost due to the long and sophisticated handle of surgical tools. To assist the surgeon to regain this perception, research in developing a tactile relay system is being conducted by many groups. Figure 11.9 shows a block diagram of a tactile sensing system.

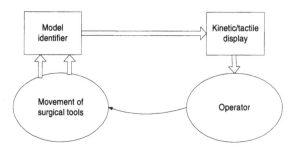

Figure 11.9. In a haptic sensing system, tool movement changes the tactile display presented to the operator.

The force-moment sensor can be a hollow cylinder-shaped deformation body attached to miniature strain gages, or an array of sensors. The haptic display located at the handle of the instrument is felt by the fingers of the surgeon.

11.3 ENDOSCOPIC ELECTROSURGERY, ULTRASONIC SURGERY AND LASER SURGERY

In addition to traditional dissection instruments, other cutting mechanisms, such as electrosurgery, ultrasonic surgery and laser surgery have been applied to endoscopic surgery. This section discusses the principle, instrumentation and advantages of these technologies.

11.3.1 Electrosurgical technologies in endoscopic surgery

Electrosurgery means cutting or coagulation of body tissue with a high-energy electric current. It is a versatile, convenient, low-cost procedure, and is easily incorporated into an endoscope. Therefore, electrosurgical technology is preferred by surgeons who are performing endoscopic procedures.

Electrosurgery uses ac current in the radio frequency range (0.5–2 MHz), because this frequency range does not cause neuromuscular stimulation, while the thermal energy is produced by the interaction between the ac current and the body tissue. In 1980, high-speed microscopic motion pictures were made to record the physical process during thermal cutting and coagulation at a cellular level. This showed that during electric cutting, fast heating of liquid in cells causes rupture of the cell membrane, while the slower heating coagulation waveform uncoils the molecular chains and then forms extended branches, which leads to a hemostatic effect.

Different waveforms are used based on the above understanding of the cutting and coagulation effect. The waveform used for cutting is a high-frequency sinusoidal waveform. Coagulation uses underdamped sinusoidal bursts allowing a short cooling period. In practice, the surgeon can select either waveform or a mixed waveform to perform the cutting and coagulating at the same time.

Depending on the application, either the monopolar or the bipolar configuration is used in electrosurgery. For example, monopolar electrosurgery effectively offers wide surface coagulating and dissection of dense tissue, while bipolar technology is applied to perform precise procedures to delicate tissue.

In the monopolar configuration, the active electrode is placed on the surgical location and the ground lead is routed to a ground plate (dispersive electrode) placed under the patient. The current passes from the active electrode, through the patient, and then returns to the generator by way of the ground plate. At the point where the active electrode makes contact with the body tissue, the current density is very high, while the current density is low at the ground plate. As a consequence, the electrosurgical effect only takes place near the active electrode.

Electric burns occurring at an alternative site are the largest potential hazard in monopolar electrosurgery. This situation might be caused by insulation failure, ground-plate contact failure, direct coupling, or capacitive coupling. Ground-plate contact failure occurs when a decrease in contact of the ground plate causes increased current density and heating. Direct coupling occurs when the active electrode touches another metal device so the current is conducted to that device, causing its adjacent tissue to be heated. Capacitive coupling occurs when current is induced in the nearby conductive material, despite intact insulation.

In a bipolar arrangement, both electrodes are placed on the patient. Current flows from one electrode through the tissue to the other electrode (figure 11.10). Bipolar electrosurgery is considered a safer technology because it virtually eliminates the risk of alternative site burns described above.

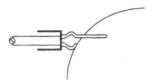

Figure 11.10. Current flows between the two electrodes in a pair of bipolar cutting forceps.

Most modern electrosurgical systems, however, incorporate both monopolar and bipolar modes in the same unit.

11.3.2 Ultrasonic surgery and harmonic scalpel

When a sound wave of sufficient intensity is applied to the tissue, small bubbles are generated. These bubbles grow if the frequency of the sound wave is high enough and the implosion of the bubbles causes tissue damage by cellular fragmentation. The dissection effect of the sound wave is most efficient in high-water-content tissue. This tissue selectivity is an advantage in endoscopic surgery because tumor tissues normally contain more water than normal tissues. Thus the unwanted tumor can be dissected with minimal damage to the surrounding normal tissue. The typical endoscopic ultrasonic surgical aspirator has a tip that oscillates at 23 kHz.

Another advanced instrument, the harmonic scalpel, operates at a higher frequency. The blade of the harmonic scalpel vibrates at 55.5 kHz in a back and forth motion. This high-speed mechanical motion results in disorganization of the tissue's protein and the formation of a sticky substance (coagulum), which seals blood vessels at the same time they are being cut. The applications of the harmonic scalpel in laparoscopy include laparoscopically assisted vaginal hysterectomy, laparoscopic bladder neck suspension (for urinary stress incontinence), and endometriosis. The harmonic scalpel probes designed for laparoscopic surgery are 30 cm long and 5 mm or 10 mm in diameter. There is a variety of tip shapes used for different purposes. The blunt tip is used to perform coagulation and the blade-shaped tip is used to cut the tissue.

11.3.3 Laser surgery

As a laser beam is absorbed by tissue, it heats the tissue as well as electric current does. Currently the CO_2 laser, argon laser and YAG laser are widely used in surgery. Table 11.1 shows their characteristics and applications.

Table 11.1. Comparison of three types of laser used in laparoscopic surgery.

	Wavelength range	Absorption by intracellullar fluid	Absorption by tissue pigment	Major effect and application
CO_2 laser	Distant infrared	High		Vaporizes intracellular fluid rapidly; cutting
Nd-YAG laser	Near infrared	Low	High	Denatures protein; coagulation
Argon laser	Visible light	Between the above two	High	Cutting and coagulation

The CO_2 laser is effective at cutting by vaporizing intracellular fluid rapidly at precise locations with relatively small amounts of injury to the surrounding tissues. Because the wavelength of the CO_2 laser cannot be propagated through conventional optical fibers, the beam is delivered by articulated or flexible hollow waveguides into the abdominal cavity. These waveguides have a narrow inner channel (< 1 mm) and bounce the beam down to the desired location.

The Nd-YAG laser and the argon laser can be propagated along optical fibers. In laparoscopic laser surgery, the optical fibers used are usually made of quartz with diameters of 600 μm, 200 μm or 100 μm.

11.4 THE BASIC PROCEDURE AND EQUIPMENT SET-UP FOR LAPAROSCOPIC SURGERY

Laparoscopic surgery is one of the most common forms of endoscopic surgery. During this procedure, the surgeon operates inside the patient's abdominal cavity by manipulating instruments through a small incision in the abdominal wall, instead of cutting a large lesion to perform the surgery as in major or open surgery.

The laparoscope was first employed to visualize abdominal disease as early as the 1900s; however, its application had been confined to only a few types of surgical procedure because of technical limitations. With the growth of emerging technologies, such as the manufacture of microcameras and microsurgical tools, along with the development of sophisticated surgical techniques, more and more major surgical procedures are being replaced by laparoscopic surgery.

This section provides knowledge of the present laparoscopic equipment and insight into the developing trends.

11.4.1 Basic procedures of laparoscopic surgery

Although the surgical procedure differs from case to case, for most types of laparoscopic surgery, the basic routines include the following steps:

1. *Insufflation.* A hollow needle is inserted through the abdomen wall, the end of which is connected to an insufflator tube. Gas is delivered into the abdominal cavity to distend the abdominal wall and separate the organs, which allows easier visualization and manipulation. The pressure inside the abdominal cavity is controlled by the insufflator to an appropriate constant value throughout the operation.
2. *Insertion of the trocar.* A trocar (cylindrical tube with a sharp tip) is inserted through the abdomen wall to introduce the laparoscope. One or more secondary trocars might also be used to bring in other instruments.
3. *Organ visualization and manipulation.* Observing and operating instruments are brought into the patient's abdominal cavity though primary and secondary trocars.
4. *Terminating the procedure.* All the trocars are taken out of the abdominal cavity and the wounds are closed.

11.4.2 Equipment set-ups for laparoscopic surgery

The equipment and tools employed in the above procedure include a laparoscope, which might be coupled with an electrosurgical unit or a laser channel, insufflator, Verress needle, trocars and other surgical tools, such as forceps and scissors.

11.4.3 Descriptions of some laparoscopic equipment and surgical tools

During laparoscopic surgery, equipment such as laparoscopes, insufflators and trocars are also designed to perform the surgery with the least trauma to healthy tissues. This section introduces some of these tools and equipment.

11.4.3.1 Laparoscopes. Laparoscopes are specially designed endoscopes to perform or assist surgical procedures inside the abdominal cavity. Typical diagnostic laparoscopes are rigid endoscopes with diameters ranging from 5 to 10 mm, although smaller, specialized laparoscopes are also available. Normally the laparoscopes equipped with a video camera have a larger diameter (8–10 mm) to guarantee adequate light thoughput. The viewing angle of a diagnostic laparoscope can be 0°, or 30–45°. Some laparoscopes are coupled with an operating channel, such as an electrosurgical channel or a laser-surgical channel. These laparoscopes normally have a larger diameter and the viewing angle is almost always 0°.

11.4.3.2 Insufflators. Before a laparoscopic surgery, pneumoperitoneum, which means delivering gas into the abdominal cavity, is performed to separate the organs, which allows easier visualization and manipulation. It is important that the intra-abdominal pressure is maintained at an appropriate value throughout the entire procedure. These requirements are achieved by a laparoscopic insufflator.

Either CO_2 or N_2O is used as the insufflation gas. Safety is the major concern in deciding which gas is to be used. This consideration includes the possibility of gas embolism, gas absorption and the possibility of provoking hypercardia with cardiorespiratory changes. Parameters such as the circulatory volume of the patient, the ventilatory technique used, the underlying pathologic conditions, and the type of anesthesia used should all be taken into account.

For either of the gases, the volume of gas introduced and intra-abdominal pressure must be carefully monitored and controlled. Most modern insufflators use a digital controller along with accurate pressure measurement system to regulate the gas pressure.

11.4.3.3 Safety trocars and Verress needles. Trocars and Verress needles are inserted into the patient's abdominal cavity to introduce the laparoscope, surgical tools or to deliver gas from the insufflator. Because the insertion progress is not under visualization, it is important to make sure that these instruments will not cause injury to the inner organs after penetrating the abdominal wall. Figure 11.11 shows the structure of a Verress needle integrated with a safety mechanism. It has a hollow outer needle with a sharp distal end, which makes it easy to penetrate the abdominal wall. An inner stylet with blunt tip is fixed on top of a spring in the handle. During penetration, the inner stylet withdraws within the outer needle because of the resistance against the abdominal wall. Once the needle tip has entered the abdominal cavity, the spring forces the stylet to move forward to an extent beyond the outer needle. This structure allows the needle to enter the body without injuring the structure inside the abdominal cavity. An indicator in the handle indicates the position of the inner stylet relative to the sharp distal end of the outer needle.

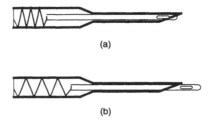

(a)

(b)

Figure 11.11. A modern Verress needle with a safety mechanism. (*a*) While penetrating, the inner stylet withdraws within the sharp outer needle. (*b*) After penetrating, the stylet extends out of the outer needle to protect the organs inside the abdominal cavity.

Trocars with similar safety mechanisms have also been designed to allow the laparoscope and other surgical instruments to enter.

11.4.4 New trends and perspectives of laparoscopic technology

Laparoscopic technology has a long history of over 80 years, beginning with diagnostic laparoscopy. However, there is still new development in this area.

11.4.4.1 Microlaparoscopy. Conventional laparoscopy performs surgery by opening incisions with a diameter of about 1 cm. Microlaparoscopy is a new technology that makes the procedure even less intrusive by making smaller incisions with a diameter of only a few millimeters.

Many micromanufacturing technologies are involved in the development of macrolaparoscopy. The employment of microfibers and microlenses has made it possible to create microlaparoscopes with a diameter of 2 mm. Small trocars (with a size of 3 mm) and tiny scissors, forceps and electrosurgical dissectors have also been produced.

11.4.4.2 Manipulating the movement of the laparoscope. Conventionally, the laparoscope is held by the surgeon or an assistant during the procedure. To improve safety, efficiency and reduce human resource utilization, robotic arms have been developed to hold the laparoscope. Such a telemanipulator is able to move the laparoscope in three axes based on the command input of the surgeon, changing the location to be observed, and zooming or enlarging the display. The command input could be foot-pedal movement, head-motion sensor, or even the voice of the surgeon. Chapter 14 provides detailed information on manipulating the laparoscope with robotic arms.

11.5 ARTHROSCOPY

Arthroscopy is a minimally invasive surgical procedure performed to diagnose and treat problems inside joints without major cuts. Similar to most endoscopic surgery, a specially designed endoscope (arthroscope) is inserted into the joint through a small incision to examine conditions inside. Other surgical tools are inserted through several other incisions to perform the surgery. The view of the surgical site obtained by the arthroscope is picked up by an attached camera and is displayed on a monitor.

Arthroscopic surgery has been applied to almost all major joints: the knee, the shoulder, the hip, the elbow, the ankle, the wrist and the lumbar spine. The conditions that can be treated with arthroscopy include:

1. treatment of arthritis and other chronic swelling caused by infections;
2. fixing or repairing injuries within the joints such as cartilage tears, ligament strains or tears, and cartilage deterioration;
3. removal of scar tissue, osteophytes (prominences that arise from the osseous margins of the joint) or foreign objects remaining inside the joints after fracture or injury.

By performing these procedures without opening the joint completely, arthroscopic surgery is much less invasive than conventional surgery and is preferred by surgeons and patients whenever it is possible.

11.5.1 Instruments

11.5.1.1 Arthroscopes. Arthroscopes are available in different sizes, focal lengths and lens angles. The typical lens diameters are 2.7 mm, 4.0 mm or 5.0 mm. Most arthroscopes used in orthopedic surgery have angles of 0, 30 or 70°.

11.5.1.2 Operable tools. A variety of specially designed manually operable tools or powered instruments are available, such as drills, probes, suture anchors, graspers and forceps, chondral awls (used to establish vascular channels in subchondral bone for examining chondral defects) and suction punches.

11.5.1.3 Lasers. Applying lasers for cutting and coagulating tissue in arthroscopic surgery becomes more and more common. To limit the damage to surrounding tissue, the type and energy settings of the laser device must be carefully determined according to tissue type, arthroscopic medium and technique of the surgeon. The first laser used in arthroscopy was a CO_2 laser. CO_2 is easily absorbed by water so it requires a gaseous medium to perform the surgery, which is hard to achieve in many arthroscopic surgery procedures. And it cannot be transmitted by fiber-optic cables and requires awkward hand-pieces. The most widely used type today is the 2.1 μm holmium:YAG laser. It is used in an aqueous medium; and it can be transmitted by fiber-optic cables; it is precise and minimizes surrounding tissue damage.

11.5.2 Arthroscopic knee surgery

An excellent example illustrating arthroscopic technology is the operation on the ACL (anterior cruciate ligament), which is located inside the knee joint.

The ACL is a ligament with interwoven fibers that connects between the femur and tibia and controls the knee's movements. The ACL might be injured when twisting movements force the knee beyond its normal range of motion. A complete tear of the ACL is the 'unraveling' of the interwoven fibers. The golden standard surgical procedure to treat complete ACL tear is ACL reconstruction, whereby the torn ligament is replaced by a tendon from the patient or by a synthetic material. The conventional surgical procedure to reconstruct the ACL involves opening the joint entirely to operate; leaving large scars and requiring a long period of recovery time. Normally the patient would wear a cast for six weeks and be immobilized for six months after this procedure.

The arthroscopic approach is much less invasive. The procedure is described below.

1. *Examination.* After setting up the surgery and general endotracheal anesthesia is administered, the arthroscope is used to survey the entire joint.
2. *Harvesting the graft.* Through an anterior incision, the patellar tendon is exposed just beneath the skin. A central strip is removed, along with attached segments of bone from the patella above and the tibia below. This is the only nonendoscopic step during the operation and requires an 80–130 mm incision.
3. *Preparing the graft sites.* With the assistance of a laparoscope, the injured ACL tissue is removed by specially designed cutting and suction devices. The intercondylar notch in the lateral wall is enlarged by drills to create a more spacious area for the new graft.
4. *Placing and then attaching the graft.* Holes with a diameter of 9 or 10 mm are made by drills into the joint through the tibia below, and then up into the femur.

5. *Placing and attaching the graft*. The graft with its sutures in each end is passed carefully through the tibial hole, the central part of the joint and the hole in the femur into the joint guided by the sutures. The graft is fixed to the bones by screws, either made of metal, or more recently of bioabsorbable materials.

6. *Closing the wound*. The small holes through which the arthroscope and instruments are inserted are closed by sutures.

After the arthroscopic procedure, patients can return home on the same day, begin rehabilitation the next day. Within two weeks the patient is able to walk without crutches and without any need for the leg immobilizer.

PROBLEMS

11.1   Explain why electrosurgical technologies are widely applied in laparoscopic surgery but not so frequently used in arthroscopy.

11.2   Give the simplest image-processing method to remove the honeybee pattern from an endoscopic image transmitted by a fiber optic bundle.

11.3   Figure 11.12 shows a conventional lens and illumination system. Compare it to the Hopkins lens system given in figure 11.1(*b*) and give at least two advantages of the Hopkins lens system.

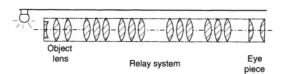

Object
lens
                        Relay system

Eye
piece

Figure 11.12. The complete optical system for a rigid endoscope.

REFERENCES

Andrews J R and Timmerman L A (ed) 1997 *Diagnostic and Operative Arthroscopy* (Philadelphia, PA: Saunders)

Boppart S A, Deutsch T F and Rattner D W 1998 Optical imaging technology in minimally invasive surgery, current status and future directions *Surg. Endosc.* **13** 718–22

Brillhart A T (ed) 1995 *Arthroscopic Laser Surgery: Clinical Applications* (New York: Springer)

Cockett W S and Cockett A T 1998 The Hopkins rod-lens system and the Storz cold light illumination system *Urology* **51** (Suppl 5A) 1–2

Dickens M M, Houlne M P, Mitra S and Bornhop D J 1997 A soft computing method for the removal of pixelation in micro-endoscopic images *Proc. SPIE* **3165** 186–94

Finlay P A and Ornstein M H 1995 Controlling the movement of a surgical laparoscope *IEEE Eng. Med. Biol.* **14** (3) 289–91

Gomel V G and Taylor P J (ed) 1995 *Diagnostic and Operative Gynecologic Laparoscopy* (St Louis, MO: Mosby)

Hunter J G and Sackier J M (ed) 1993 *Minimally Invasive Surgery* (New York: McGraw Hill)

LaCourse J R, Rothwell A D and Selikowitz S M 1993 Development of electrosurgery. I: A historical perspective *Proc. 1993 IEEE 19th Annual Northeast Bioeng. Conf.* (New York: IEEE) pp 35–6

Linder T E, Simmen D and Stool S E 1997 Revolutionary inventions in the twentieth century. The history of endoscopy *Arch. Otolaryngol. Head Neck Surg.* **123** 1161–3

Soderstrom R M (ed) 1998 *Operative Laparoscopy* 2nd edn (Philadelphia, PA: Lippincott-Raven)

Tomkinson T H, Bentley J L, Crawford M K, Harkrider C J, Moore D T and Rouke J L 1996 Rigid endoscopic relay systems: a comparative study *Appl. Opt.* **35** 6674–83

Vasin L N, Kletskaya M A, Antonova I V, Frolov Y A, Yagmurov V K and Potemkin A I 1994 Very thin flexible endoscope for urology *J. Opt. Technol.* **61** 912–14 (Translated from *Optiko-Mekhanicheskaya-Promyshlennost.* **61** 67–9)

CHAPTER 12

IMAGE-GUIDED SURGERY

Oliver Wieben

Visualization in conventional surgery is somewhat limited, because the surgeon cannot see beyond exposed surfaces. This issue becomes even more crucial in minimally invasive surgeries, where the entry into the body should be kept as small as possible. Therefore, it is desirable to have some form of image guidance that allows the surgeon to control the position of surgical devices and to expand his/her view beyond surfaces to comprehend complex anatomical structures.

A variety of medical imaging modalities are available to support diagnosis and surgical planning. For example, X-ray computed tomography images have exquisite characteristics for imaging bone structures; magnetic resonance imaging has good soft tissue contrast; X-ray digital subtraction angiography and MR angiography can depict blood vessels; and PET and functional MRI show functional information of brain or organ activity. All imaging techniques have their specific strengths and weaknesses and often it is desirable to combine information from multiple anatomical and functional imaging modalities. In addition, other data sources such as EEGs and MEGs can be registered with anatomical images. The surgeon may benefit from computer-assisted image fusion and 3D displays to ease the mental challenge of combining information with varying resolutions and orientations. Advanced image-processing algorithms and computer graphics can register and fuse data from multiple sources so that regions such as target tissues and vital tissues can be identified.

Image guidance can be based on preoperative images combined with localization methods or on images acquired during the procedure, e.g. intraoperative X-ray, ultrasound, or MR imaging. The use of images acquired prior to the surgery requires tracking of surgical instruments in the operational field and mapping of the images to the surgical field. These images can serve as roadmaps during the surgical procedure to add information to the limited view of the surgeon and to guide surgical instruments to the desired locations. However, the anatomy of a patient may shift after image acquisition, especially during a surgical procedure when tissue is removed and instruments are inserted into the body. Alternatively, the anatomy and the surgical devices can be imaged in real-time or near to real-time by intraoperative imaging methods.

This chapter describes the key components for image-guided minimally invasive surgeries: registration techniques, surgical planning, frame-based and frameless stereotactic systems, and imaging modalities that allow for intraoperative imaging in real-time. The described methods are used, for example, for head surgeries, hip replacements, spinal surgeries, biopsies, and drainage of cysts.

12.1 IMAGE REGISTRATION

The spatial alignment of image data from sources with differences in spatial resolution, field of view, orientation, scale, patient position, or anatomical contrast is referred to as image registration. Once the images are registered (their spatial relationships to each other are known), they can be reformatted on a common coordinate system. Since image data are represented by 3D volume elements of finite size (voxels), the fusion and display on one grid may require interpolation techniques. Terms such as registration, coregistration, fusion, matching and integration are found in the literature to describe these processes. Image registration can be used for the alignment of images from multiple imaging techniques (multimodal imaging) or from time series of images from the same technique (monomodal imaging). The registered data can originate from a single patient (intrasubject registration), from two patients or a patient and a model (intersubject registration), or a patient and a generic atlas. The algorithms vary in the complexity of allowed transformations between images, the techniques to derive the transformations, the degrees of user interaction, their speed, and their accuracy. Comprehensive reviews on image registration can be found elsewhere (Maurer and Fitzpatrick 1993, van den Elsen *et al* 1993, Maintz and Viergever 1998).

12.1.1 Rigid body transformation

Image registration can be expressed as a transform or mapping of a point $x = (x,y,z)^T$ in one space to the same point in another space or coordinate system $x' = (x',y',z')^T$, where an upper case T denotes the transposed vector. This transform can either be globally applied on the complete imaged scene or it can vary locally. The type of motion that can occur between the original and the remapped coordinate system is restricted by the mathematical model for the transformation. The most basic transformation describes rigid bodies and allows only for rotation and translation. The rigid body mapping function f can be described in matrix notation as

$$f(x) = Ax + b \tag{12.1}$$

where A is a rotation matrix (3×3) and b a translation vector (3×1). Alternatively, the coordinate transformation can be expressed by a single transformation matrix F:

$$X' = F X \tag{12.2}$$
$$X = F^{-1} X' \tag{12.3}$$

where F^{-1} is the inverse of F. In this representation, the four-dimensional vectors $X = (x,y,z,1)^T$ and $X' = (x,y,z,1)^T$ replace x and x' so that a single 4×4 transformation matrix

$$F = \begin{pmatrix} F_{11} & F_{12} & F_{13} & F_{14} \\ F_{21} & F_{22} & F_{23} & F_{24} \\ F_{31} & F_{32} & F_{33} & F_{34} \\ 0 & 0 & 0 & 1 \end{pmatrix} \tag{12.4}$$

describes the geometric transform sufficiently. In this matrix, the upper left 3×3 submatrix is a rotation matrix and the upper right 3×1 submatrix describes a translation vector (Lemieux *et al* 1993). Only six parameters are required to determine F. Rescaling operations can also be included in rigid body transformations and figure 12.1 shows an example for such an image registration in two dimensions.

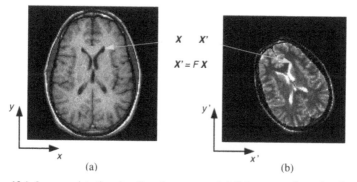

(a)   (b)

Figure 12.1. Image registration describes the process of defining a transformation that can map points from one space to another. Images (a) and (b) show two differently weighted MR images of the same patient with changed positions and acquisition parameters. A rigid body transformation matrix F can map a point X from the coordinate system in (a) to the according point X' in coordinate system (b). With the transformation matrix a point in (a) can be directly related to a point in (b), although the coordinate systems are rotated, translated and scaled.

12.1.2 Nonrigid body transformation

Image registration based on global rigid body coordinate transformation found wide-spread use in head imaging. However, it cannot compensate for local variations such as tissue deformations or geometric distortions in acquired images. Solutions for these problems include local transformations, subdivision of the image into smaller regions, and more complex transformations. If two or more transformations are applied to the same image or subimage, then this is referred to as a local transformation. They have to satisfy local continuity restraints and are used sparely. More often, a rigid coordinate transformation is applied to a section of the image, where the global transformation holds for the selected region.

Affine transformations are restricted to rotation, translation and scaling but also allow for shearing. Shearing can be interpreted as a nonorthonormal scaling operation on one or more axes. Affine transformations map parallel lines into parallel lines and can be performed using matrix multiplications as used for rigid transformations.

More general but also more complex models are described by elastic (or curved) transformations. They allow for deformations that obey constraints such as continuity or smoothness. Elastic transformations are frequently represented by vector displacement fields because no constant transformation matrices exist. Only very few algorithms have been applied in clinical practice because of the large parameter space for the solution. Figure 12.2 illustrates rigid, affine and elastic transformations on a simple example.

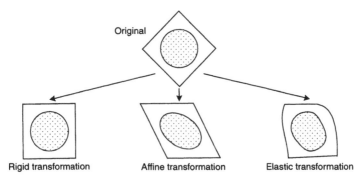

Figure 12.2. Examples for typical transformations used in image registration. Rigid transformations include rotation, translation and possibly scaling. Affine transformations expand on rigid transformations and incorporate shearing also. Elastic transformations are based on smoothness or continuity constraints and are very complex to model.

12.1.3 Extrinsic image registration

Image registration algorithms must determine the transformation as accurately as possible. Many registration techniques require the presence of foreign objects in the image scene to support precise registration. These techniques are referred to as extrinsic or prospective registration approaches because stereotactic frames or other markers have to be attached to the patient before any images are acquired. If markers are used, they can be fixed rigidly to the patient or can be mobile (fixed to soft tissue). Rigid markers require a minor surgical procedure but are more accurate as compared to mobile markers. The base of rigid markers is typically made of plastic and implanted in a bony structure. Fiducial markers with contrast materials for MRI, CT or PET can be mounted on the bases.

Extrinsic image registration is mostly used with rigid body transformations. The registration process requires the identification of common points (at least three noncollinear points) before the volume can be registered. The fiducial points require good contrast in all data sets so that they can be identified by automated algorithms. The transformation matrix can be found, for example, by minimizing the mean difference of all markers in the two data sets. Section 12.3 discusses the use of frame-based and frameless systems in stereotactic surgeries and presents an example for such an image registration. The algorithms are typically fast and robust because the registration is based on the alignment of a few points. The accuracy of these methods depends on the precision of the fiducial points.

12.1.4 Intrinsic image registration

Intrinsic registration techniques are based on the image content without fiducial reference systems. They are also called retrospective because they do not require any precautions at the time of image acquisition. The algorithms fall into three categories: (1) voxel property based, (2) landmark based, or (3) segmentation based. West *et al* (1997) evaluated the execution time and accuracy of 12 multimodal registration algorithms.

12.1.4.1 Voxel property-based methods. Voxel property-based methods use the gray-scale values of the imaged scene or scalars derived from them. Measures such as cross-correlation, squared intensity difference, intensity variance, image ratio uniformity and mutual information are used to assess the similarity between images. An iterative process can realign the data sets with an optimization procedure until a mapping within a predefined error margin is found. In the first step of a representative algorithm (Woods *et al* 1998), data are interpolated to cubic voxel dimensions. The data sets are divided into reference images X_{ref} and reslice images X. The reslice images are resampled on the grid of the reference image with the current parameters of the spatial transformation ($X' = F\,X$). If no scaling between the data sets is assumed, then the spatial transformation consists of three angular and three linear offsets between the image sets. A cost function evaluates the similarity between the reference and reslice images. An optimization procedure computes new spatial transformation parameters and iteratively minimizes the cost function. Once the cost function is smaller than a predefined threshold, the procedure terminates.

Such techniques work particularly well when the data sets originate from the same imaging modality, for example time-separated MRI data sets. Their use for multimodal images is limited because of the differences in contrast characteristics from one modality to another. Voxel property-based algorithms can achieve subvoxel accuracy (the average registration misalignment is smaller than the voxel dimension) but require large computational efforts.

Two objects can also be registered by aligning their centroids and principal axes. In this case, measures that have their analog in mechanics are assigned to the imaged objects based on their signal intensities. A set of three orthogonal principal axes is inherent to each three-dimensional solid body and these axes can be found and aligned for the different data sets. In this case, the information in the gray-scale values of an imaged scene is compressed to scalar quantities.

12.1.4.2 Landmark-based methods. Instead of extrinsic fiducial points, anatomical or geometric landmarks can be identified for a point-based registration. Anatomical landmarks such as vessel bifurcations are typically interactively defined. The use of anatomical landmarks has proven to be labor intensive and inaccurate because of the operator dependence. Alternatively, automated algorithms can identify geometric landmarks such as corners or local extremes in curvatures. Once the landmarks are found, the registration procedure is equivalent to the methods described for extrinsic registration.

12.1.4.3 Segmentation-based methods. Image segmentation can be used to automatically extract curves or surfaces from salient objects. If these curves or surfaces are identified in both image sets, then rigid body transformations and even elastic transformations can be found because of the availability of many common points. For example, the contours of the imaged object can be extracted for surface-based image registration. The distance between these contours can then be minimized by automated optimization procedures. Examples of extracted curves are meshes that include ridge and crest lines. All of these methods are highly dependent on the accuracy of the segmentation algorithm.

12.1.5 Image fusion

Once image registration determines the spatial relationship between image volumes, they can be reformatted on a common coordinate system. Figure 12.3 shows data sets from multiple imaging sources that have been registered on a common grid. Since image data are represented by 3D volume elements of finite size (voxels), the resampling on a new grid requires an interpolation technique such as trilinear or sinc-interpolation. Isotropic voxel dimensions are advantageous for the accuracy of the interpolation algorithms.

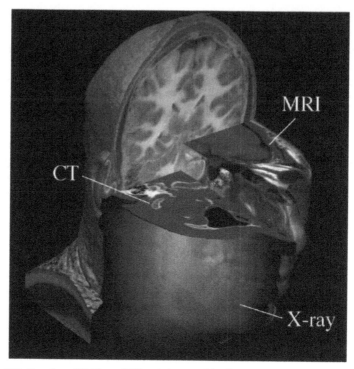

Figure 12.3. Data from CT, X-ray, MRI and photographic slices are shown side by side after their registration on a mutual coordinate system (courtesy of K H Hoehne, Institute of Mathematics and Computer Science in Medicine (IMDM), University of Hamburg, Germany).

12.2 SURGICAL PLANNING

Preoperative surgical planning is important for minimally invasive surgeries. In neurosurgery, the identification of the target areas, such as a tumor, and vital tissues, such as larger blood vessels or essential parts of the brain, allow the determination of an optimal surgical corridor for the intervention. Measurements of the anatomy can be taken prior to surgery to optimize the shape of implants such as hip prostheses or stent grafts. Surgical planning can be based on patient-specific data only or with the incorporation of generic models (e.g. brain atlas). Recent advances in image processing, computer graphics and visualization allow for the fusion of information and sophisticated three-dimensional representations of complex data.

12.2.1 Generic atlas models

In addition to patient-specific data, generic atlases can be used as additional sources of information. Multiple groups in the world have compiled atlases based on measurements of thousands of patients, especially brain atlases (St-Jean *et al* 1998). Analogous to the image registration procedure, the task becomes to match a standard atlas to images obtained from an individual (Vannier and Marsh 1996). A commonly used brain coordinate system for atlases is the Talairach proportional grid system, which uses some anatomical landmarks. In addition to geometric transformations, elastic deformation may be used to accomplish the goal. An obvious problem of all approaches is the difference in anatomy from patient to patient and even for the same patient, e.g. from one hemisphere to the other. Mapping with data from generic atlases becomes extremely challenging for individuals with pathologies or mass lesions.

12.2.2 Visualization

The visualization of image-based information is a nontrivial task. The goal is to present large amounts of data (most often 3D data) in a highly efficient manner. A simple approach is reformatting of the data sets and their display in three orthogonal body axes (axial, sagittal and coronal slices as shown in figure 9.2 and figure 12.7). However, this still requires the surgeon to mentally generate a 3D model from multiple 2D images.

The computer revolution had a big impact on the graphic engines of modern workstations. Today it is possible to generate interactive 3D visualizations with off-the-shelf hardware and software components (e.g. ANALYZE, Mayo Clinic, Rochester, MN). 3D representations such as surface-shaded display and volume rendering allow for interactive examinations of data from different angles and magnifications. Figure 12.4 shows an example of a volume-rendered scene for the planning of brain surgery.

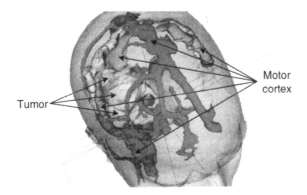

Figure 12.4. Advanced visualization methods support surgical planning and allow for interactive examination of the acquired data in three dimensions. In this scene, the motor cortex and an identified tumor are displayed to assist in surgery planning. Color displays allow for better separation of the identified regions (adapted from Grimson *et al* 1999).

The identification of relevant anatomic structures requires image segmentation algorithms (see chapter 10) and often manual image processing (Kikinis *et al* 1996). The resulting three-dimensional information can be used for orthopedic planning (e.g. optimal prosthesis placement), optimized shapes of stent grafts for the individual patients in vascular surgery (Shahidi *et al* 1998), etc. Representation of multiple imaging modalities in one model requires image mapping, fusion and visual clues on the different information contained. This can be accomplished, for example, by different color tables or by selective visualization. With present technology it is even possible to generate stereoscopic 3D displays and holograms that allow 3D exploration of the data. Chapter 13 describes some of these techniques in more detail.

12.3 STEREOTACTIC SURGERIES

Stereotactic systems have been used since the turn of the century for anatomical studies and surgical procedures of the brain. The term is derived from the Greek root *stereo* (meaning three-dimensional) and the Latin root *tactic* (meaning to touch). It refers to the localization of structures in the body (Lemieux *et al* 1993). Stereotactic systems establish a coordinate system that can map pre- or interoperative images (image space), surgical instruments and the surgical field (physical space) all in one coordinate system to accurately guide surgical procedures. Methods available to perform the registration of image space and physical space include (1) atlas methods, (2) curve and surface methods, and (3) point methods. The main applications of stereotactic imaging are head surgeries (including neurosurgery, sinus surgery, oral surgery and maxillofacial surgery), hip surgeries and spinal surgeries. High spatial fidelity for the registration is imperative for these procedures and two categories for proper registration exist: frame-based and frameless stereotactic surgery. Frame-based stereotactic systems require an external frame that is attached to the patient prior to image acquisition. The methods discussed below differ in their accuracy, the physical principles used to derive positions in the surgical field, and their intrinsic advantages and disadvantages.

12.3.1 Frame-based stereotactic systems

In frame-based stereotaxis, an external frame is attached to the patient prior to the acquisition of pre-operative data. After fixing the frame under local anesthesia, a diagnostic scan, e.g. with CT or MRI, is performed. Target areas (e.g. a cyst or tumor) and important anatomy such as blood vessels and specific brain regions can be identified to find optimal trajectories for the surgical procedure. The frame is also used as a basis for mounting surgical tools and, therefore, providing a well-defined coordinate system, e.g. with an arch on top of the frame. The frame is visible in the images and can be used for the registration of multimodal images.

The preoperative images have to be registered with the frame coordinate system with a transformation matrix F. The frames are designed so that they allow for the derivation of this matrix. Among the most popular frame designs is the Brown–Roberts–Wells (BRW) system, which mainly consists of three N-shaped structures. Figure 12.5 shows a stereotactic frame with N-shaped localizers.

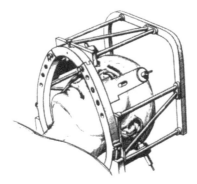

Figure 12.5. A stereotactic frame for neurosurgery. The frame is fixed to the patient and three N-shaped structures surround the imaged volume (from Kelly 1993).

The N-shaped assemblies consist of three rods each. The N-shaped structures allow the determination of the height of an image with respect to a reference point and the orientation of an image slice. Figure 12.6 shows the principle of the N-shaped arrangement.

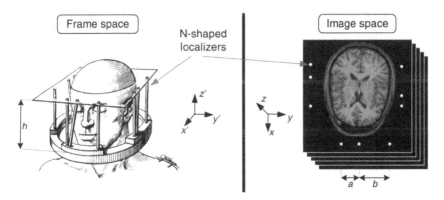

Figure 12.6. N-frames for the registration of image space and frame space. The height h above the base rim of the frame can be derived from measurements a and b in image space: $h \propto a/(a + b)$. The registration of image space and frame space requires the identification of at least three noncollinear points in both coordinate systems (adopted from Kelly 1993).

The task is to map the acquired images from image space X to the frame coordinate system X' (equation (12.2)). If we select a slice in image space we can ignore the z-component in image space. We can then formulate

$$
\begin{pmatrix} X \\ Y \\ Z \end{pmatrix} = F \begin{pmatrix} X' \\ Y' \\ 1 \end{pmatrix}
\tag{12.5}
$$

with F being a 3×3 matrix (the third column and the fourth row of the 4×4 matrix can be removed here).

For proper registration it is sufficient to find three noncollinear points within the N-shaped assembly whose locations are known in image space $X_i = (X_i, Y_i, 1)^T$ and in the frame coordinate system $X_i = (X'_i, Y'_i, Z'_i)^T$. For example, two matrices $U = (X'_1 X'_2 X'_3)$ and $V = (X_1 X_2 X_3)$ can be constructed from these three points. Since $U = F V$, the matrix F can be easily found as

$$
F = U V^{-1}
\tag{12.6}
$$

where V^{-1} denotes the inverse matrix of V. If more than three points are used for the registration, the system becomes overdetermined and generally more than one solution exists. The best solution can be found with a minimization norm such as the least square solution (Lemieux *et al* 1993).

Other aspects of frame design include mechanical precision, good contrast and easy adjustment of the surgical tools to the target point. In reality, accuracy and frame flexibility have to be compromised. The advantage of frame-based systems is their precision (about 2 mm) while their disadvantages include the invasive attachment, the cumbersome use, and the missing position feedback on preoperative images. In addition, only images that have been acquired with the attached frame can be used. Despite their disadvantages, frame-based systems are utilized very often for head surgeries and stereotactic biopsies (Kucharczyk and Bernstein 1997).

12.3.2 Frameless stereotactic systems

Frameless stereotactic systems link pre- or intraoperative images, the operative field, and the location and orientation of surgical instruments in one coordinate system. Spatial information on surgical devices can then be displayed, e.g. on the images with sets of cross hairs or with augmented and virtual reality techniques. The computer functions as a localizer and requires the establishment of a geometric relationship of the images to the operative field and tracking of the surgical instruments. Different tracking methods have been proposed including optical digitizers (infrared optical probes), electromagnetic trackers, ultrasonic sensor assemblies, and electromechanical digitizers (Anon *et al* 1997). These techniques are also referred to as interactive, image-guided surgery techniques.

12.3.2.1 Infrared optical probes. Surgical navigation systems based on infrared optical probes require an infrared optical digitizer and light-emitting diodes (LEDs) mounted, for example, on surgical instruments. Figure 12.7 shows such a set-up in the operating room. The position of the patient can be localized with LEDS that are mounted on a rigid head holder or alternatively on a head band for dynamic tracking.

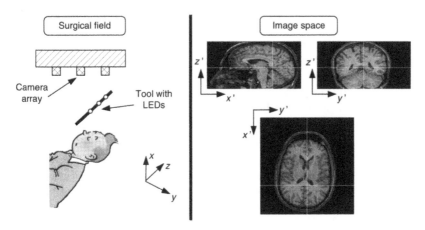

Figure 12.7. Intraoperative navigation based on infrared optical digitization and a camera array. The position of surgical tools in the surgical field can be determined from LED signals that are recorded by a camera array. If the patient position is known also then the tools can be displayed on preoperative images.

The light emitted from the LEDs is detected by three charge-coupled devices (CCDs), which are aligned so that the fields of view of the cameras are perpendicular to each other (Bucholz and Greco 1996). The cameras are mounted above the surgical field and the infrared light is focused by cylindrical lenses in front of each CCD camera (typically 2048 elements). Which of the CCD elements are illuminated depends on the spatial location of the firing LED. The actual position of the LED in a 3D coordinate system can be determined by triangulation. For the establishment of a reference coordinate system of the surgical field it is sufficient to identify three LED positions (see section 12.1.3). The LEDs never emit light simultaneously and thus the point in time identifies which LED is firing.

Any image-guided surgical procedure starts with the registration of the image data with the surgical field. This intraoperative registration can be accomplished with fiducial markers that are visible in the images and the operative field and therefore can establish the transformation matrix. The fiducials are fixed to the patient and represent fixed reference points in both coordinate systems. Alternatively, anatomical landmarks can be mapped with a position-sensing pointing device that is physically held to the position of the landmark and also identified in image space, e.g. with the placement of a cursor. Point-based registration methods can be used for the mapping. Yet another approach is surface-based methods, where 40 or more fiducial markers are attached to the skin. The mapping can be performed with surfaces extracted from the images and fiducials without the need for a pointing device. If the patient is moved, the coordinate systems have to be reregistered.

LEDs are also mounted on the surgical instruments to allow for their tracking during the procedure. Since optical coupling requires a free line of sight between the LED and the camera, the LED cannot be fixed on the tip of the surgical instrument. One solution to this problem is the attachment of at least two LEDs in a linear arrangement with the tip. With knowledge of the distances from these LEDS to the tip and of the spatial location of the LEDs (with the CCDs), the position of the tip can be calculated. Typically, a third LED is added for redundancy in case one of the LEDs is not visible on a camera.

Image-guided systems based on infrared optical probes are very accurate (nominal accuracy of 0.2–0.3 mm without registration errors) and commercial products are capable of tracking up to 256 LEDs. The disadvantage of these systems is the requirement of a free optical path between the LEDs and the CCD cameras. Obstructions of LEDs lead to incomplete coordinate acquisition. This requirement may become limiting in a surgical field with occupied staff, intensive care units and various surgical instruments.

12.3.2.2 Electromagnetic trackers. Commercial products exist for electromagnetic digitizers also (e.g. InstaTrak System from Visualization Technology, Wilmington, MA). These devices are capable of detecting gradients in magnetic fields, which are generated by source coils in a known fashion. Three-dimensional positions and orientation (pitch, yaw and roll) can be determined from voltages induced in coils in the sensor. The sensors are connected to the image-guidance system with cables. They do not require a free line-of-sight as do the infrared optical probes, but they are very dependent on a stable magnetic field.

The magnetic field can be generated by alternating currents (ac) or pulsed direct currents (dc). The alternating currents typically have a frequency of 8–14 kHz. In this case, three coils are used as sensors to measure the induced voltage. In dc systems, three flux gate sensors measure the components of the magnetic induction in three dimensions. The presence of ferromagnetic objects (various metals, surgical tools, ultrasonic scan probes, etc) in the operating room leads to field distortions and thus inaccuracies in both systems. The accuracy of electromagnetic position digitizers is approximately 3–6 mm (nominal accuracy of 1.0 mm and 0.15°), depending on registration accuracy. The accuracy can be improved if the magnetic field is carefully calibrated and nonferromagnetic instruments are used (Birkfellner *et al* 1998).

12.3.2.3 Ultrasonic sensor assemblies. The use of ultrasonic (US) sensor assemblies is another approach for interactive image localization. The systems are similar to the optical digitizers based on infrared probes. Here, US emitters are used for the tracking. A microphone array of known geometry records the lag time between the emission of ultrasound waves and their arrival at the microphones. The position of the emitter can be calculated from these measurements. As for LED probes, the system relies on a free line of sight. The speed of sound varies with temperature, which results in accuracy errors if the temperature varies from the position of the probe to the microphone (Anon *et al* 1997).

12.3.2.4 Articulated mechanical arms. The use of electromechanical digitizers with a mechanical arm provides another method for image-guided surgery. One of the most popular products is the ISG Viewing Wand (ISG Technologies Inc., Missiauga, Ontario, Canada). The system combines a real-time visualization package with a six degree-of-freedom passive mechanical arm. The sensing arm can be mounted on a stereotactic frame or the operation room table. Electropotentiometers sense the position of the arm at six joints and analog-to-digital converters feed this information to the processing unit. Once the patient space and image space are registered, the tip of the arm can be localized because the geometry of the arm and its probe tip is known (McDermot and Gutin 1996). A typical update rate for the position is 30 Hz. The calculated position can then be displayed on the preoperative images and assist the surgeon in navigation.

Passive arms are steered by the surgeon while active arms are steered towards the treatment area by computer control. Active arms may make dangerous moves if any component in the control software or the image registration fails.

The use of articulated arms requires rigid fixation of the patient. Therefore, the surgeon must find a single position for the patient throughout the surgical procedure. Otherwise he has to reregister when moving the patient for better access, for example, or when using fluoroscopy. They are also limited in their accuracy (2–5 mm). On the other hand, they do not require a free line-of-sight.

12.4 INTRAOPERATIVE ENDOSOCOPY AND MICROSOCOPY

All image-guidance methods discussed so far assume that the preoperatively acquired images represent the anatomy sufficiently during the surgery. Unfortunately, these premises are not always met. Shifts and changes in the anatomy may occur due to loss of cerebrospinal fluid (CSF), hemorrhage, edema, the influence of gravity in position changes of the patient and last but not least the movement and removal of tissues during surgery. These problems can only be overcome with intraoperative imaging where images are acquired during the procedure. Therefore, the imaging modality must be present in the operation room to allow for identification of the present position of structures of concern and the relative position of surgical instruments (Galloway *et al* 1993). Stereotactic and intraoperative guidance do not compete but can complement each other.

12.4.1 Endoscopy

The use of endoscopes produces keyhole images from the interior of the body. The 2D image is spatially limited by anatomy surfaces and the angle of acceptance of the lens. During some procedures, disorientation may occur. Where in the body is the surface that is displayed on the endoscopic display? What is the rotation of the image? Even highly trained experts may have problems because of the length of a rigid endoscope, varying viewing directions on the tip, the limited view, and difficulties in finding landmarks due to disease or injuries. For example, navigation through the nasal cavity for sinus surgery is difficult because it is narrow and its inner structure is complicated. Disorientation may lead to serious surgical errors such as damage to the optical nerve (Yamashita *et al* 1999).

The use of image guidance can assist the surgeon in navigating through complex structures. The tip of the endoscope and its orientation can be tracked, for example, with an electromagnetic sensor or a mechanical arm as described in section 12.3.2 (Fried and Morrison 1998). The position of the lens or its field of view can be displayed as cross hairs in orthogonal images. This display allows the surgeon to guide the endoscope with information on the surrounding anatomy and its precise location. In addition, concepts from virtual endoscopy (as described in chapter 13) can be incorporated. With this technique, the surgeon is exposed to a 3D representation of the anatomy based on the preoperative images. Different from the 2D view on the endoscope display, selected anatomical regions can be removed interactively to look behind surfaces (Yamashita *et al* 1999). This display may improve the orientation during the interventional procedure.

12.4.2 Microscopy

Just as it is possible to track the position of surgical tools and endoscopes, microscopes for neurosurgical applications can be located. The tracking can be based on LEDs (Bucholz and Greco 1996), ultrasonic digitizers (Roberts *et al* 1993), or mechanical systems (Roessler *et al* 1998). Knowledge of the position of the microscope tip and its focal length allows for the display of the focal point on the preoperative images. The multicoordinate manipulator system (MKM) from Zeiss (Oberkochen, Germany) was developed as a carrier for surgical microscopes and is shown in figure 12.8. The target area can be identified during surgery planning and its distance and location can be displayed during the intervention. Contours of the target area can be superimposed on the micrsocope images. The movement of the carrier of the surgical microscope is controlled by foot, joystick (hand or mouth) or voice control to free the hands of the surgeon. Roessler *et al* (1998) report an application accuracy of 2.2 mm with this system after a learning period.

Figure 12.8. The multicoordinate manipulator system (MKM) from Zeiss. The carrier of the microscope can be controlled by foot, hand, mouth or voice. Information on target areas identified during surgery planning can be superimposed onto other images in the microscope and on the screen.

12.5 X-RAY FLUOROSCOPY

X-ray fluoroscopy allows us to image dynamic processes in real-time (Bushberg *et al* 1994). Besides diagnostic applications such as the barium swallow exam and digital subtraction angiography (DSA) as shown in figure 12.9, fluoroscopy is used for interventional procedures. One of the most common interventional procedures is percutaneous transluminal balloon angioplasty (see chapter 10). Other procedures include RF ablation and other treatment of vascular disease. These interventions are performed with a catheter that is commonly inserted in the patient's femoral artery. The catheter is then advanced to the area of interest where the procedure is guided by 2D projection images in real-time (30 frames s^{-1}). For biopsies the position of a needle is tracked. The images are displayed on a monitor in the operation room and the surgeon can adjust image acquisition with a joystick and foot pedals.

   X-ray-based image guidance offers exquisite spatial resolution and good delineation of bone structures. The main disadvantages include the exposure of the patient and the staff to the primary X-ray beam and scattered radiation. X-ray images also lack good soft tissue contrast.

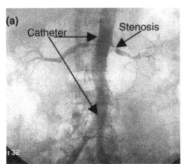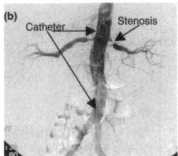

Figure 12.9. A digital subtraction angiography (DSA) exam of the abdomen. A catheter is inserted through the femoral artery and brought with a guide-wire into the abdominal aorta. (*a*) A contrast agent is injected through the catheter and the arterial blood vessels become visible under X-ray fluoroscopy due to the opaqueness of the contrast medium. (*b*) The subtraction of a precontrast mask eliminates static signal and improves the vessel delineation. This patient has a stenosis of the left renal artery. Fluoroscopy images have high temporal resolution and high spatial resolution in two dimensions (courtesy of R Omary, University of Wisconsin–Madison).

Fluoroscopy is widely used and offers the advantages of X-ray-based images and a high temporal resolution. The images are projections through the patients and sometimes two fluoroscopy units with perpendicular axes are used to reveal more spatial information. The main disadvantage of fluoroscopy is the use of ionizing radiation. The described procedures can be fairly long and the dose can reach critical levels. Potentially, high X-ray exposures can cause severe skin damage and cancer.

12.6 INTRAOPERATIVE COMPUTED TOMOGRAPHY

CT guidance allows for the acquisition of 3D data sets based on axial slices. The image scene is represented in 3D rather than in 2D projection images as in fluoroscopy. It is used for minimally invasive biopsies, e.g. providing depth information of a tumor for the guidance of a transthoracic needle biopsy (Klein and Zarka 1997). It also allows more complex needle trajectories compared to fluoroscopy, such as angled approaches. Figure 12.10 shows one slice from a CT-guided biopsy of a lymph node. The procedure starts with an initial CT scan and the identification of the suspicious lymph node. In an iterative process, the needle is advanced several times while the location is verified with subsequent CT scans. Another application of CT guidance is the drainage of cysts. It is even possible to evaluate the success of the drainages by intraoperative measurements of the cyst volume (Kondziolka and Lunsford 1993).

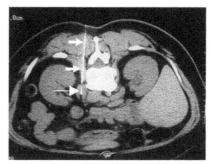

Figure 12.10. CT-guided biopsy of a lymph node. A 56-year-old female with known cervical carcinoma had a 2 cm × 3 cm left para-aortic lymph node (bottom arrow) suspicious for metastatic disease. Under CT guidance, with the patient prone, a 22 gauge fine needle aspiration biopsy of the lymph node was performed. The image shows a slice in which the needle (top two arrows) was advanced into the lymph node. Cytology indicated metastatic squamous cell carcinoma (courtesy of R Omary, University of Wisconsin–Madison).

CT images have high spatial resolution in all three dimensions and the tip of a needle is accurately visualized. However, data acquisition and display are not possible in real-time. Therefore, CT-guided exams take quite a long time and require cooperative patients. CT guidance also requires a high dose of X rays.

12.7 INTRAOPERATIVE ULTRASOUND

Ultrasound (US) is an imaging modality that combines various advantages: it is portable, noninvasive, displays images in real-time, can encode blood velocities, and is of low cost (see chapter 10). These features and the ease of use of the small US probes make it a frequently used tool for image-guided procedures, e.g. biopsies, drainages and ablations. With US, the site of interest can be continually monitored during the intervention.

Disadvantages of US include the reduced signal-to-noise ratio (SNR) and a reduced spatial resolution compared to other imaging modalities. The field of view is somewhat limited and the image quality varies with depth in the patient. US images are acquired in 2D planes, which can be registered and assembled into a 3D volume (3D US).

12.8 INTRAOPERATIVE MAGNETIC RESONANCE IMAGING

Magnetic resonance (MR) imaging for guidance in the operating room is appealing for many reasons. MR imaging does not require ionizing radiation, has high soft tissue contrast, allows for reconstruction of images in any desired plane, and can be sensitized for a variety of parameters such as blood flow, tissue perfusion, functional brain mapping and temperature mapping. However, performing surgeries under MR guidance also brings about many challenges. The scanner design must allow for access to the patient and potentially for instrument tracking. Surgical instruments and the equipment in the operating room have to be MR compatible. The imaging protocol must allow for fast imaging, and the data have to be reconstructed in near to real-time.

12.8.1 Scanner design

In conventional MR scanners, the patient is placed inside a tubelike magnet bore. The scanners use superconducting magnets and produce strong magnetic fields (typically 1.5 T) but the patient is inaccessible to the surgeon. In order to use MR imaging for therapy guidance, the conventional design has to be modified. Magnets with shorter bores can be used for interventional procedures that are based on the use of catheters. The manufacturers of MR scanners proposed different solutions for scanner design with direct access to the imaging field (open magnets). Besides permitting better access to the patient, the open scanners are also more patient friendly and allow for scanning of claustrophobic patients.

Siemens (Siemens AG, Erlangen, Germany) promotes an open magnet with a field strength of 0.2 T (magnetom open scanner). The field is aligned vertically by two separated resistive electromagnets (no superconducting wiring) as shown in figure 12.11(a). The horizontal gap is approximately 44 cm and the C-shaped assembly permits a 260° patient access. The patient table can be easily attached and removed. This scanner type can be used for biopsies and some percutaneous procedures. However, it is not suitable for percutaneous and open surgical procedures that require hand–eye coordination and close and direct contact with the anatomic regions of interest (Jolesz 1997). Steinmeier *et al* (1998) report the use of this system for surgically treated cerebral lesions. MR images were acquired and the patient was moved to a conventional operating theatre with conventional equipment and unlimited patient access. If necessary, the patient was brought into the MR imaging room again for updating the images.

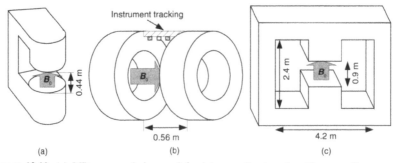

Figure 12.11. (*a*) MR scanner design used for intraoperative imaging. Siemens offers an open scanner (0.2 T) with resistive electromagnets, (*b*) GE Medical systems designed the doughnut 'double-doughnut' system specifically for intraoperative imaging (0.5 T), and (*c*) Fonar proposed an open scanner with 360° access to the patient (0.6 T).

GE Medical Systems (Milwaukee, WI) and researchers at the Brigham and Women's Hospital in Boston, MA developed a superconducting 0.5 T open magnet specifically for intraoperative MR imaging (Black *et al* 1997, Grimson *et al* 1999). The magnetic field is created by a Helmholtz coil assembly and aligned horizontally as in conventional scanners. Due to its geometry it is known as the 'double doughnut' magnet design. Figures 12.11(*b*) and figure 12.12 show how the surgeon has access to the patient through a 56 cm gap in between the field generating coils.

Figure 12.12. Image-guided neurosurgery with an intraoperative MR unit at the Brigham and Women's hospital in Boston, MA. The surgeon has access to the patient through a small gap between the magnet windings. A live video image of the patient is fused with a 3D computer model from presurgical scans and displayed on a monitor. Subsequent MR scans provide information on tissue displacements and deformations (from Grimson *et al* 1999).

Fonar (Fonar™ Corp., Melville, NY) is currently working on a scanner design with 360° patient access and a patient gap of 90 cm (OR 360). One resistive electromagnet is installed on the floor and the other on a standard 2.40 m ceiling to create a 0.6 T horizontal magnetic field as shown in figure 12.11(*c*). The room has a maximum width of 4.20 m to feed back the magnetic flux and can be of any desired length.

The magnet wires are made of niobium tin, which is superconducting at higher temperatures than conventional windings. Therefore, liquid helium is not necessary around the winding for cooling purposes and the gap is wider. The gradient coils are also installed in the double-doughnut assembly (Lamb and Gedroyc 1997). Conventional transmit and receive RF coils can be used with the system, but preclude access because they typically surround the patient. Specifically designed flexible surface coils are available and can be attached to the patient with Velcro straps. Sensors needed for tracking information can be placed in the same gap above the patient. This configuration can also be used for motion studies when the patient is standing in between the two rings. With this design, the patient is directly accessible, but the space for the surgical team (surgeons and nurses) is very limited.

12.8.2 Instrumentation compatibility

Most of the traditionally used surgical equipment and instrumentation has to be modified when used in a MR operating theatre. Only nonmagnetic metals should be present in the MR room and RF pulses from, for example, electrosurgical tools may cause severe artifacts. Surgical tools, endoscopic equipment, etc have to be MR compatible. Otherwise, they may not only distort the images, but they may also become dangerous projectiles in the strong magnetic field. Anesthesia machines and patient monitors have to specifically designed so that they do not malfunction in the magnetic field or cause any artifacts or distortions. Images and other information cannot be displayed on conventional monitors. Liquid crystal display based monitors are used instead.

Surgical tools that are within the imaging field should have a magnetic susceptibility similar to that of human tissue (Gould and Darzi 1997). Otherwise, these objects cause distortions in the reconstructed images. The magnetic susceptibility of ceramics, nylon and plastics is very close to that of human tissues, but only a few surgical instruments can be manufactured with these materials. Nonmagnetic materials such as titanium, tungsten, nitinol and aluminum can be used but cause artifacts in the images. Instruments are commonly made out of titanium/vanadium alloys, which produce acceptable artifacts.

12.8.3 Instrument tracking

In addition to the tracking methods described above, active or passive MR tracking can be used for the localization of the surgical instruments. In active tracking, small RF coils are implemented in the surgical instruments such as catheters, biopsy needles, endoscopes, etc. The position of the coil is frequency encoded by the application of gradient pulses. The frequency recorded in the coil is analyzed and registered with the displayed MR images. The location of the coil can then be displayed in the image data.

Alternatively, passive tracking methods can be used for localizing instruments in the MR image data. Materials that introduce displacement of the magnetic spins can be used for the instruments. The susceptibility artifact causes signal void in the image. Recently, the use of contrast materials to label and track instruments has been investigated. Contrast agents similar to those used in MR angiography can be chemically bound, for example to catheter coatings, and thereby enhance large parts of the object.

12.8.4 Data acquisition and reconstruction

The goal of real-time or near to real-time imaging requires fast imaging protocols. Currently used imaging sequences include fast gradient echo sequences, fast spin echo sequences, and echo planar imaging (EPI). Recent advances in gradient and coil design allow for faster imaging and increased SNR and this trend is most likely to continue. The reconstruction of MR data sets is typically based on Fourier transforms (FTs). If data are constantly acquired and potentially interpolated, the image reconstruction via 3D FTs becomes computationally fairly demanding. High-performance computer hardware is necessary to reconstruct and display the 3D data in a near to real-time fashion.

PROBLEMS

12.1 Discuss methods to register images from multiple modalities.
12.2 Explain the role of image registration in surgery planning and interactive image-guided surgery.
12.3 Define the terms frame-based and frameless stereotaxis.
12.4 Explain how image space and frame space can be mapped with a stereotactic frame.
12.5 List methods to perform frameless stereotactic surgery.
12.6 Explain the principle of navigation with optical digitizers.
12.7 List and compare intraoperative imaging techniques.
12.8 Explain the advantages and challenges of intraoperative imaging.
12.9 Explain required modification to the MR unit to allow for interactive image-guided surgery.

REFERENCES

Anon J B, Klimek L, Mosges R and Zinreich S J 1997 Computer-assisted endoscopic sinus surgery — an international review *Otolaryngol. Clin. N. Am.* **30** 389–401

Birkfellner W, Watzinger F, Wanschitz F, Enislidis G, Kollmann C, Rafolt D and Nowotny R 1998 Systematic distortions in magnetic position digitizers *Med. Phys.* **25** 2242–8

Black P M, Moriarty T, Alexander E, Stieg P, Woodard E J, Gleason P L, Martin C H, Kikinis R, Schwartz R B and Jolesz F A 1997 Development and implementation of intraoperative magnetic resonance imaging and its neurosurgical applications *Neurosurg.* **41** 831–42

Bucholz R D and Greco D J 1996 Image-guided surgical techniques for infections and trauma of the central nervous system *Neurosurg. Clin. N. Am.* **7** 187–200

Bushberg J T, Seibert J A, Leidholdt E M Jr and Boone J M 1994 *The Essential Physics of Medical Imaging* (Baltimore, MA: Williams & Wilkins)

Dean D, Kamath J, Duerk J L and Ganz E 1998 Validation of object-induced MR distortion correction for frameless stereotactic neurosurgery *IEEE Trans. Med. Imaging* **17** 810–16

Fried M P and Morrison P R 1998 Computer-augmented endoscopic sinus surgery *Otolaryngol. Clin. N. Am.* **31** 331–40

Galloway R L Jr, Beger M S, Bas W A and Maciunas R J 1993 Registered intraoperative information: electrophysiology, ultrasound, and endoscopy *Interactive Image-Guided Neurosurgery* ed R J Maciunas (Park Ridge, IL: American Association of Neurological Surgeons)

Gould S W T and Darzi A 1997 The interventional magnetic resonance unit—the minimal access operating theatre of the future? *Br. J. Radiol.* **70** S89–97

Grimson W E L, Kikinis R, Jolesz F A and Black P M 1999 Image-guided surgery *Sci. Am.* **280** (6) 62–9

Jolesz F A 1997 Image-guided procedures and the operating room of the future *Radiol.* **204** 601–12

Kelly P J 1993 Stereotactic excision of brain tumors *Stereotactic and Image Directed Surgery of Brain Tumours* ed D G T Thomas (Edinburgh: Churchill Livingstone)

Kikinis R, Gleason P L, Moriarty T M, Moore M R, Alexander E, Stieg P E, Matsumae M, Lorensen W E, Cline H E, Black P M and Jolesz F A 1996 Computer-assisted interactive three-dimensional planning for neurosurgical procedures *Neurosurg.* **38** 640–49

Klein J S and Zarka M A 1997 Transthoracic needle biopsy: an overview *J. Thoracic Imaging* **12** 232–49

Kondziolka D and Lunsford L D 1993 Intraoperative computed tomography localization *Interactive Image-Guided Neurosurgery* ed R J Maciunas (Park Ridge, IL: American Association of Neurological Surgeons)

Kucharczyk W and Bernstein M 1997 Do the benefits of image guidance in neurosurgery justify the costs? From stereotaxy to intraoperative MR *Am. J. Neuroradiol.* **18** 1855–9

Lamb G M and Gedroyc W M W 1997 Interventional magnetic resonance imaging *Br. J. Radiol.* **70** S81–8

Lemieux L, Henri C J, Wootton R, Collins D L and Peters T M 1993 The mathematics of sterotactic localization *Stereotactic and Image Directed Surgery of Brain Tumours* ed D G T Thomas (Edinburgh: Churchill Livingstone)

Maciunas R J, Berger M S, Copeland B, Mayberg M R, Selker R and Allen G S 1996 A technique for interactive image-guided neurosurgical intervention in primary brain tumors *Neurosurg. Clin. N. Am.* **7** 245–66

Maintz J B A and Viergever M A 1998 A survey of medical image registration *Med. Image Analysis* **2** 1–36

Maurer C R Jr and Fitzpatrick J M 1993 A review of medical image registration *Interactive Image-Guided Neurosurgery* ed R J Maciunas (Park Ridge, IL: American Association of Neurological Surgeons)

McDermott M W and Gutin P H 1996 Image-guided surgery for skull base neoplasms using the ISG viewing wand — anatomic and technical considerations *Neurosurg. Clin. N. Am.* **7** 285–95

Roberts D W, Friets E M, Strohbehn J W and Nakajima T 1993 The sonic digitizing microscope *Interactive Image-Guided Neurosurgery* ed R J Maciunas (Park Ridge, IL: American Association of Neurological Surgeons)

Roessler K, Czech T, Dietrich W, Ungersboeck K, Nasel C, Hainfellner J A and Koos W T 1998 Frameless stereotactic-directed tissue sampling during surgery of suspected low-grade gliomas to avoid histological undergrading *Minimally Invasive Neurosurg.* **41** 183–6

Shahidi R, Tombropoulos R and Grzeszczuk R P 1998 Clinical applications of three-dimensional rendering of medical data sets *Proc. IEEE* **86** 555–68

Steinmeier R, Fahlbusch R, Ganslandt O, Nimsky C, Buchfelder M, Kaus M, Heigl T, Lenz G, Kuth R and Huk W 1998 Intraoperative magnetic resonance imaging with the magnetom open scanner: concepts, neurosurgical indications, and procedures: a preliminary report *Neurosurg.* **43** 739–47

St-Jean P, Sadikot A F, Collins L, Clonda D, Kasrai R, Evans A C and Peters T M 1998 Automated atlas integration and interactive three-dimensional visualization tools for planning and guidance in functional neurosurgery *IEEE Trans. Med. Imaging* **17** 672–80

van den Elsen P, Pol E J D and Viergever M A 1993 Medical image matching—a review with classification *IEEE Eng. Med. Biol.* **12** 26–39

Vannier M W and Marsh J L 1996 Three-dimensional imaging, surgical planning, and image-guided therapy *Radiol. Clin. N. Am.* **34** 545–63

West J *et al* 1997 Comparison and evaluation of retrospective intermodality brain image registration techniques *J. Comput. Assisted Tomography* **21** 554–66

Woods R P, Grafton S T, Holmes C J, Cherry S R and Mazziotta J C 1998 Automated image registration: I. General methods and intrasubject, intramodality validation *J. Comput. Assisted Tomography* **22** 139–52

Yamashita J, Yamauchi Y, Mochimaru M, Fukui Y and Yokoyama K 1999 Real-time 3D model-based navigation system for endoscopic paranasal sinus surgery *IEEE Trans. Biomed. Eng.* **46** 107–16

CHAPTER 13

VIRTUAL AND AUGMENTED REALITY IN MEDICINE

Oliver Wieben

Virtual reality (VR) is an emerging technology that could drastically change traditional medical training and surgical techniques. Virtual environments create, for example, three-dimensional visual and sound perception and tactile feedback. The level of interactivity in real-time distinguishes VR applications from other computer systems.

In medical applications, users can interact with such virtual worlds to study complex anatomy; assist in surgical planning; perform minimally invasive and image-guided surgery; and perform surgical simulation. It may even be possible to perform surgeries on physically remote patients with data links to receive sensory information from a specially equipped operating theatre and transmit control signals to a surgical robot. Some of these applications may sound like part of a science fiction novel, but most are already used in research laboratories around the world and some systems are commercially available. Where will the future lead us? In order to play a significant role in medicine, VR systems must be realistic (especially in visual and tactile fidelity) with short latencies, validated, affordable, and beneficial to patients. The rapid progress in computer technology makes predictions difficult, but these systems may revolutionize some aspects of traditional health care.

The virtual reality applications presented in this chapter are meant to be examples rather than a complete list of all available systems. Many more prototypes and commercial products exist and are continuously added to this rapidly developing area.

13.1 VIRTUAL ENVIRONMENT

In virtual environments (VE), users are *immersed* in artificially generated environments. They can experience their new surroundings in two ways:

receiving information by using their senses and interacting with so-called actuators. Therefore, a virtual reality system must generate sensory inputs for the human and also sense the action of the user and information from the environment to create the illusion of an alternative reality. Figure 13.1 illustrates the concept of VE. The user receives input from the real and the virtual environment through the senses (sight, hearing, touch, taste and smell). The user can act on his environment with body motion, including eyes, hands, mouth and facial expressions. The computer system uses sensors to record changes in the real environment and inputs from the user. The system can also interact with the user and the real environment using special actuators, which are described in this chapter (Sharma *et al* 1998). Other terms for VE include cyberspace, telepresence, mirror world, artificial reality, wraparound compuvision and synthetic environments (Dumay 1996). A realistic experience of a VE requires real-time processing with a delay between input and system response smaller than 10–100 ms. This poses high demands on all system components.

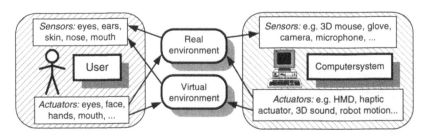

Figure 13.1. Schematic diagram of a virtual environment (VE). The computer generates a VE based on sensory input from the real environment. Specially designed actuators allow the computer to act on the real and virtual environment. The user can interact with both environments. (HMD = head-mounted display.)

13.1.1 VR sensors

13.1.1.1 Movement and position sensors. The most traditional computer-input devices for hand movements are the keyboard, mouse, joystick, light pen and tablet. Spaceballs, joysticks and 3D mice have been used for positioning in three dimensions and can determine the position with six degrees of freedom (x, y, z, azimuth, pitch and roll). However, for virtual reality applications it can be important to track position and motion of head, fingers and other body parts. Such tracking may be used for the recognition of hand gestures or the control of devices. The methods discussed in section 12.2 can also be applied for VE applications: electromagnetic sensors, light emitting diodes (LEDs) and ultrasound probes. In addition, inertial tracking based on accelerometers can be used. The most commonly used probes for head tracking are electromagnetic sensors, although they can be used over a limited area only.

The tracking of hand movements is complicated, because the hand has more than 20 degrees of freedom. One approach for hand tracking is the use of thin gloves which directly measure hand angles with engraved optical fibers (Greenleaf 1996). The optical fibers are connected to a LED and a photodetector. The detected light attenuates with increasing flexion of the fiber. A variety of commercial products exist with varying numbers of fibers, precision and update rates. Figure 13.2(*a*) shows the *CyberGlove*® from Virtual Technologies, Inc. (Palo Alto, CA), one of the most popular models. This glove is available with 22 sensors (three flexion sensors for each finger, four abduction sensors, a palm-arch sensor, and sensors to measure wrist flexion and abduction). Alternatively, a glove with 18 sensors has open fingertips to allow better grasp and typing. The update rate for the glove is 150 records per second if all sensors are recorded. The manufacturer reports a sensor resolution of 0.5° and a 1° sensor repeatability between glove wearings. Such gloves can be equipped with spatial trackers to determine location and orientation of the hand. However, it typically requires many additional cables and can be somewhat cumbersome.

Figure 13.2. Input devices for virtual environments. (*a*) The CyberGlove® for the tracking of hand movements (image courtesy of Virtual Technologies, Inc., Palo Alto, CA, USA). Finger flexion is monitored with fiber optics and a position sensor can be attached to track the position and orientation of the hand in space. (*b*) The STAR*TRAK® system captures position and motion of the whole body with an electromagnetic system (image courtesy of Polhemus Inc., Colchester, VT, USA). (*c*) The Gypsy™ 2.5 Motion Capture System analyzes whole body motion on the basis of potentiometers (image courtesy of Analogus™ Corp., Fremont, CA, USA).

There are also commercial systems for tracking whole body motion in 3D space. For example, the STAR*TRAK® from Polhemus Inc. (Colchester, VT, USA) is a real-time RF motion capture system based on electromagnetic sensors that can operate wireless. Figure 13.2(*b*) shows how up to 32 receivers with an update rate of 120 Hz each can be attached to two performers. The capture range of this system is up to 6 m between transmitter and receiver and the accuracy in absolute space is ± 2.5 cm and ± 2° per receiver. Alternatively, body motion can be captured with optical and electromechanical systems. The *Gypsy*^{TM} *2.5 Motion Capture System* from Analogus^{TM} Corp. (Fremont, CA) is a suit based on potentiometers and torque transmitting sliding bars. It captures movements of the head, clevical, upper and lower arms, hands, spine, hips, upper and lower legs, and feet (heel and ball). The system is shown in figure 13.2(*c*). The Gypsy 2.5 weighs approximately 10 kg with cables and has a sampling rate of up to 120 Hz. The manufacturer reports a resolution of 0.125° and a wireless option with a range up to 300 m is available. The system can track body motion inside and outside but not the position in absolute space.

13.1.1.2 Visual sensors. Visual information can be acquired using video systems. The analysis of image sequences allows for the tracking of hand gestures, lip movements, gaze, facial expressions, and head, body, and even eye movements (Sharma *et al* 1998). The described tasks require high-performance computing because the acquired amounts of data are large. This is especially true under real-time conditions. However, computer vision and image processing are areas of active research and commercial products are available. These systems use multiple cameras or just one camera mounted on the head to capture facial expressions. The attachment of fiducial markers to cloth or skin is often required for tracking of movements or facial expressions.

13.1.1.3 Audio sensors. Sound waves, particularly the voice, can be recorded with microphones. The ultimate goal is the automated recognition of human speech even under noisy conditions and with more than one speaker. Some systems for voice recognition have been developed but their robustness must improve to work reliably in clinical settings.

13.1.1.4 Tactile and force sensors. An essential component of human interaction with the environment is the exertion of force. VR systems can sense forces and touch with haptic devices. This ability allows the user to touch, move, squeeze, etc objects in the artificial environment. Accurate sensing of force remains a topic of ongoing research and haptic interfaces are described in the next section.

13.1.2 VR actuators

13.1.2.1 Visual actuators. The human senses react to specific inputs, called stimuli. Although all five human senses receive stimuli, they contribute with different weights to the perception of an environment. The primary human sense is sight and about 70% of the perceived incoming information is related to vision (Faulkner and Krauss 1996). In order to be realistic, a VE must include sophisticated displays, which are ideally updated 30 to 60 times per second. Standard computer monitors display 2D information, which do not generate realistic representations of the 3D world.

A more suitable display is achieved with liquid crystal display (LCD) shutter glasses. These glasses can create 3D illusions from 2D displays. The computer generates left and right eye images sequentially (multiplexing) and the glasses alternately block and pass light for each eye (demultiplexing). The stereoscopic image display is synchronized with the glasses over a wire or an infrared signal. Therefore, the right lens is transparent and the left lens is opaque when the image for the right eye is displayed. The images alternate so fast (80–160 views per second) that the user perceives a 3D impression of the scene. Figure 13.3(*a*) shows the *CrystalEyes*®, a popular product from StereoGraphics Corp. (San Rafael, CA). These shutter glasses are wireless and weigh as much as normal eyeglasses (less than 100 g). The images can be generated on normal computer monitors with modified graphic cards and the emitter has a range up to 6 m.

a   b

Figure 13.3. Eyewear for visual display of virtual environments. The *CrystalEyes*® in (*a*) are shutter glasses that alternately block and pass light for each eye and are used with stereoscopic displays. Synchronization of display and glasses is achieved with infrared signals from the emitter shown to the left (image courtesy of Stereographics Corp., San Rafael, CA). The ProView™ XL50 in (*b*) is a head-mounted display that can deliver stereographic full color images at high resolution (image courtesy of Kaiser Electro-Optics Inc., Carlsbad, CA).

Head-mounted displays (HMDs) can also provide 3D visual feedback to the user. HMDs are worn as a helmet and generate stereoscopic images on two screens (one for the left eye and one for the right eye). They can use LCDs, small cathode ray tubes (CRT), or even optical fibers for the generation of the images. Special optics are required to bring the images in focus for the individual user. Commercial and research products differ in their display types, maximum resolution, maximum field of view (FOV), weight, etc. It is important for some medical applications that the display is not distorted due to interference from electrosurgical tools. Figure 13.3(*b*) shows the ProView™ XL50 HMD from Kaiser Electro-Optics, Inc. (Carlsbad, CA). The model series has a full color, active matrix thin film transistor (TFT) display and can provide a 1024×768 extended graphics array (XGA) resolution with a diagonal FOV of up to 70°. The user can see visual information from both the real world and the virtual environment with some HMDs, as will be discussed in section 13.1.3. Users of HMDs sometimes report some nausea due to the lag in image updates in response to head movements. There are also similar 3D virtual binoculars (e.g. from n-Vision, Inc., McLean, VA) that can be used, for example, to simulate the surgeon's view through a microscope in ophthalmic surgery.

Alternatively, 3D worlds can be displayed when shutter eyewear is combined with larger rear-projection screens. In this case, the field of view can be expanded to a large screen or even a room and many users can be immersed in the same VE. Such systems require motion tracking of the hand a user interfaces, such as 3D pointers. The ImmersaDesk R2 from Fakespace Systems (Mountain View, CA) is shown as an example of a large screen system in figure 13.4(*a*). The screen measures 2.09 m in the diagonal and the typical resolution is 1280×1024. Fakespace Systems also promotes the *CAVE*™ in which projections on three walls and the floor are generated in a 3 m × 3 m × 3 m cubic environment. Figure 13.4(*b*) shows the set-up of the CAVE™ in which one person can be tracked and up to six viewers can be immersed simultaneously.

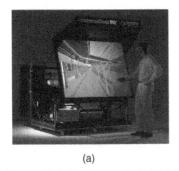

(a)   (b)

Figure 13.4. The ImmersaDesk R2 (*a*) and the CAVE™ (*b*) use shutter glasses and large rear projectors to immerse one or more users into a virtual environment The display is adjusted to the tracking of head movements and the user can interact with the environment, for example with 3D input devices (images courtesy of Fakespace Systems, Mountain View, CA).

The display of graphical information is very crucial for user acceptance. Small inconsistencies in the display can lead to significant confusion. The processing of image information in real time with sophisticated representations of objects such as surface rendering and volume rendering is a challenge for the computer system. Large improvements have been accomplished recently with the development of new algorithms, hardware and parallel computing (Soferman *et al* 1998). Another approach for 3D visualization is holography. Holograms cannot be used interactively, but can help to visualize complex data structures.

13.1.2.2 Audio actuators. The perception of sound also facilitates the feeling of immersion within a virtual environment. About 20% of perceived information is related to sound waves. Recently, 3D spatial sound systems have been developed that greatly enhance the perception of sound information. Commercially available products for 3D audio are offered from Aureal Semiconductor (Fremont, CA) and VSI Visual Synthesis (Fairfax, VA).

13.1.2.3 Tactile and force actuators. Approximately 4% of the information from all sensory input originates from sensory receptors of the skin. This includes the sensation of touch, texture and temperature of object surfaces, and applied forces such as resistance, weight and pressure. A haptic feedback system can provide tactile and/or force feedback and must be updated more than 100 times per second for a realistic impression. Burdea (1996) presents a summary of VR interfaces related to haptic sensations.

Sensations of heat can be generated by small electric heat pumps but heat actuators are seldom used. Figure 13.5 shows some commercial haptic systems that mainly provide hand force feedback and resistance to movement. Virtual Technologies, Inc. (Palo Alto, CA) offers the *CyberTouch*™ and the *CyberGrasp*™ in combination with the *CyberGlove*®, which has been described above. The *CyberTouch*™ uses six small vibrotactile stimulators (one on each finger and one on the palm), which can generate variable strength touch sensations. It enables the user to experience pulses, sustained vibrations and customized actuation functions when interacting with virtual objects. Another potential use is the 'visualization' of nonobjects such as ground-density data, magnetic field strength, etc, which can be encoded proportional to the vibrational intensity. The vibrational amplitude can be up to 1.2 N peak-to-peak at a frequency of 0–125 Hz. The *CyberGrasp*™ features one actuator for each finger and one for the palm which provide resistive force feedback by dc motors. Grasp forces of up to 12 N per finger are produced by a network of tendons routed to the fingertips via an exoskeleton. It allows full range of motion of the hand and has a workspace radius of 1 m. The *Phantom* product line from SensAble Technologies, Inc. (Cambridge, MA) consists of desktop devices for haptic feedback (Salisbury and Srinivasan 1997). The system can sense position with six degrees of freedom and a nominal positional resolution of 0.02 mm. Force feedback is generated with three degrees of freedom and a maximum force of 22 N.

Figure 13.5. Systems for haptic feedback. (*a*) CyberTouch™ provides vibrotactile stimulators on each finger and the palm of the hand (image courtesy of Virtual Technologies, Inc., Palo Alto, CA). (*b*) CyberGrasp™ features resistive force feedback for the hand from dc motors (image courtesy of Virtual Technologies, Inc., Palo Alto, CA). (*c*) The Phantom Premium 3D systems are high-fidelity desktop devices that can generate high forces (image courtesy of SensAble Technologies, Inc., Cambridge MA).

13.1.2.4 Other senses. The sense of smell (olfaction) contributes 5% to all sensory information. So far, the ability to differentiate scents has not received much attention in VE. The sensation of taste (gustation) plays a minor role in the sensory input system of the human being (1%) and no role in VR applications up to date.

13.1.3 Augmented reality

Augmented reality describes the combination of a view of the real world with computer-generated objects. Such a concept is used, for example, for the display of otherwise hidden information during surgery (Wagner *et al* 1997). The real surgical environment is displayed and additional information providing visual access to anatomy, physiology and function is superimposed. There are different implementations of such a system: a HMD can be constructed as an optical see-through or as a video see-through. Figure 13.6 shows both concepts (Tang *et al* 1998). An optical combiner uses a semitransparent mirror for the optical see-through HMD. As a result, the real environment is perceived simultaneously with the computer-generated objects. For video see-through HMDs, a video camera is attached on top of the HMD. The video images are shown with the superimposed computer graphics. Both approaches require precise registration of the surgical field with the HMD and the preoperative images. Commercial HMD products of both types are available.

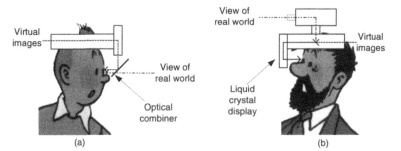

Figure 13.6. An optical see-through HMD (*a*) and a video see-through HMD (*b*). Both HMDs are used for augmented reality systems to combine images from the real world and computer-generated objects.

13.2 TEACHING

Virtual environments can serve as a basis for teaching in medical science. Visually and spatially complex topics such as anatomy, physiology, biochemistry and molecular biology can be explored using VE in ways not possible in the real world (Hoffman and Vu 1997). Objects of interest can be modified, e.g. in size or opacity. Endoscopic views can be generated for internal structures and additional information such as labels can be accessible to any desired extent.

Generic models can be created from data sets such as the *Visible Human* project, which contains high-resolution data from a CT scan, an MRI scan, and photographic slices from a male and a female individual (Ackerman 1998). Generic data sets can be used to establish models of organs or the whole body, which in turn can be explored in an interactive environment. Thus, the need for studies on animals and cadavers may be reduced. In addition, small structures such as the temporal bone in the inner ear can be enlarged for better understanding (The Virtual Reality in Medicine Lab 1999). This and some other virtual reality teaching modules support 4D processing (3D space and time).

13.3 DIAGNOSIS AND SURGICAL PLANNING

An ideal VE for patient care would integrate all available information for a patient into one model, including medical images, biochemical analysis and patient history. This model could serve as a platform for diagnosis, surgical planning, operative rehearsal and simulation, image guidance during surgery, and postoperative evaluation (Jones *et al* 1998). In this ideal scenario, multiple users could interact with the environment in real time and experience realistic impressions of sight, sound and touch. The model would be accessible even by remote physicians with telemedicine. This ideal VE does not yet exist despite dramatic improvements towards more realism and interactivity in the last decade. The computational requirements for segmentation, registration, fusion, enhancement and photorealistic stereoscopic displays of 3D image data sets in real time are high. In addition, real-time feedback for the other human senses is required for a complete immersion into the artificial environment. The following sections show applications and solutions for some of these issues.

13.3.1 Diagnosis

Patient-specific data can be used individually or in combination with generic data sets such as a brain atlas. Information from multimodal sources for diagnosis may be more intuitively displayed and accessed in VE. Recently, a variety of virtual endoscopy techniques were developed including bronchsocopy, colonoscopy, angioscopy and otoscopy (Neri *et al* 1999). Endoscopic procedures are used for the internal examination of body cavities and are somewhat invasive and often unpleasant for the patient. The motivation for virtual endoscopy is the use of a noninvasive imaging technique (e.g. CT or MRI) for the generation of a computer model. The surgeon can then perform the endoscopic procedure in the model rather than on the patient. Parts of the body that previously have been too dangerous (e.g. inside the eye) or too small (e.g. inner ears) to reach can be examined with the virtual endoscope. The field of view is not limited in the virtual anatomy by optical imperfections or the finite flexibility of the real endoscope. Additional global views and corresponding image slices can be displayed for improved orientation. In addition, information not seen in traditional endoscopy is accessible. Figure 13.7 shows examples of various virtual endoscopies derived from the Visible Human dataset at the Biomedical Imaging Resource (Mayo Foundation/Clinic, Rochester, MN).

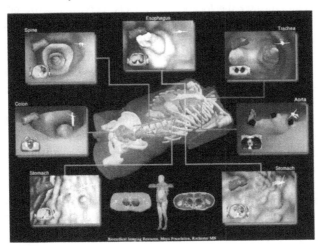

Figure 13.7. Examples for virtual endoscopies. The physician can fly through the anatomy and look at images with a similar angle as obtained from a real endoscope without the patient discomfort and risks. He can also look behind surfaces and receive detailed information on the current position and orientation (figure provided courtesy of Dr Richard A Robb, Biomedical Imaging Resource, Mayo Foundation/Clinic, Rochester, MN).

Virtual angiography generates the impression of a flight through the vascular system without the interfering presence of blood. Vascular malformations, narrowings, etc can be detected easily. In virtual otoscopy, an endoscopic procedure to cover the full extent of the ear does not exist because of the small dimensions of the inner ear. In this case, virtual otoscopy can generate views that would not be available otherwise (Frankenthaler *et al* 1998). Images similar to endoscopic video images are created by visualization techniques such as surface rendering with simulated lighting and shadowing.

Despite the advances in computer graphics the resolution of virtual endoscopy is not as good as a video image from a real endoscopic procedure (Soferman *et al* 1998). In addition, this technique is dependent on the presence of good diagnostic images. Noisy data may lead to misclassification in the segmentation process for the surface-rendered display and may cause a false diagnosis.

13.3.2 Surgical planning

In surgical planning, the surgeon reviews the accessible patient information and develops a plan for the surgical procedure. As discussed earlier, advanced visualization techniques can support the mental task of understanding complex anatomical data. Potentially, additional information from different examinations or generic data sets can be combined for the virtual patient model. Based on such a model, optimal corridors, for example neurosurgery, can be identified. These trajectories can consider the location of vessels, nerves, specific brain regions and other critical anatomy, and landmarks can be set for them when image-guided surgery is used.

Figure 13.8 illustrates how visualization can be further enhanced by the use of VR output devices for 3D rendered images. The Kent Ridge Digital Labs (Singapore) developed the Dextroscope™, a dextrous, reach-in, high-resolution 3D environment for medical applications. The Dextroscope™ is used, for example, for virtual intracranial visualization and navigation (VIVIAN) in preoperative neurosurgical planning and simulation (Serra *et al* 1998). Presurgical data are fused and preprocessed so that they are displayed as one 3D virtual object. Colors and transparencies of anatomical structures such as blood vessels, ventricles and the skull base can be changed with virtual buttons. Stereoscopic images are generated on a mirror so that the user can reach behind the screen and into the scene to manipulate virtual objects with surgical tools such as a virtual drill and suction device.

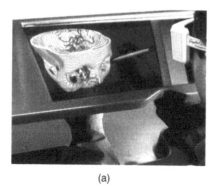
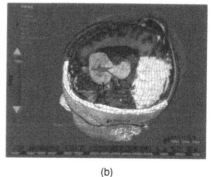

(a)   (b)

Figure 13.8. The Dectroscope™ creates a virtual environment for surgical planning and simulation. Presurgical data are fused and a stereoscopic display is generated on a mirror (*a*). Surgical tools are tracked and the user can modify the displayed scene with virtual buttons (*b*). Up to three users can view the scene and, for example, discuss and simulate surgical strategies (images courtesy of Kent Ridge Digital Labs, Singapore).

Robb (1996) describes the virtual reality assisted surgery program (VRASP) developed by the Biomedical Imaging Resource at Mayo Foundation/Clinic (Rochester, MN). The clinical goal of the system is dynamic fusion of the preoperative data with the patient in the operating room with access to the presurgical plan. The system is intended for craniofacial, orthopedic, neurologic and urologic surgery. For the planning of prostatectomy, the prostate gland, bladder, membranous urethra, rectum, contiguous muscle and the pelvic girdle are displayed in a VE. This visualization requires image preprocessing, segmentation, surface detection, feature extraction and polygon surface tiling for fast interactive display. Understanding the patient-specific anatomy may reduce the morbidity for this complicated surgical procedure. Another application for VE systems in surgery planning is the measurements for vascular or ligament stents to optimize the shape of implants (Rosen *et al* 1996).

13.4 VR SIMULATIONS

The use of simulations is one of the most appealing applications for VR in medicine. The success of flight simulators for pilot training shows how they can assist in learning difficult procedures and life-threatening situations in the real world. VR simulations can be used for training and certification of surgeons, especially in complicated procedures such as in minimally invasive medicine. Instead of practicing on animals or cadavers, the trainees can perform surgeries in the VE and repeat the procedure as often as desired. New techniques can be developed and tested in the VE and the outcome of surgical procedures can be simulated on patient-specific data. The availability of medical simulators for teaching of surgeons and medical students will likely be a very important application of virtual reality in medicine.

Surgical simulators consist of three components: computer, physical model and interface. Physical models of the patient, instruments and the operating room are required to mimic their behavior during the simulation. If the physician cuts with a scalpel, he wants to see the tissues reacting according to their elasticity, surface tension, etc, and he wants to feel the resistance of tissues with a haptic interface. Unfortunately, the calculations of physical interactions with structures such as the human body are very complex and require extensive computations.

13.4.1 Surgical simulation

With high-performance surgical simulators the physicians learn faster, procedures have a higher degree of precision, reliability, safety and better cost efficiency. Residency times are predicted to decrease and physicians can perfect their skills before operating on their first patient. The simulators can be used for certification purposes to evaluate skills and competence objectively. New procedures can be practiced in the virtual environment. Also, the use of animal experiments for education, training and testing can be limited.

Satava and Jones (1998) define four levels of simulators: (1) needle-type simulators that have only basic visual objects and a minimal axis of a single haptic device (e.g. used for spinal tap, needle insertion, central venous catheter placement, liver biopsy); (2) catheter/scope-type simulators in which the view on a video monitor has to be synchronized with the position of a control handle or catheter; (3) task-oriented simulations with one or two instruments (e.g. cross clamping, ligate and divide, anastomoses); and (4) full operative procedures. Realistic implementations of the lower level simulators exist, but the higher level simulators are of smaller visual and haptic fidelity.

Figure 13.9(*a*) shows a computer-generated model of a lower extremity. A total of 41 muscle–tendon complexes are simulated to predict the outcome of hip joint replacement. The user can interactively move tendons and the hip prosthesis in varying locations. The maximum force exerted by muscles can be calculated for varying leg movements and tendon lengths (Delp and Zajac 1993). An endoscopic sinus surgical simulator was introduced by Edmond *et al* (1998). They used the *Visible Human* data set for the generation of a virtual patient. The user holds replicas of an endoscope and a forceps in his hands as shown in figure 13.9(*b*). Their position is tracked by a six degree-of-freedom electromechanical tracking system and images are synthesized on a screen according to the position of the instruments. This system provides haptic feedback for the challenging navigation of the instruments through the rigid sinus region. Dissection is simulated in a simplified way to avoid the difficulties with complex material interaction. Marescaux *et al* (1998) developed a hepatic surgery simulator. The virtual liver was also derived from the Visible Human data set and artificial spherical tumors could be manually inserted into the model. The simulator was developed to provide a comprehensive visualization, to allow surgeons to practice surgical procedures, and to make surgeons familiar with the virtual reality concept for real surgeries. The liver can be manipulated and deformed in real time with a laparoscopic tool (Immersion Corp., San Jose, CA) as shown in figure 13.9(*c*). The displacement of the tool has an accuracy of 25 μm and the tissue deformations are shown on a computer monitor.

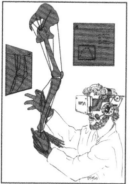 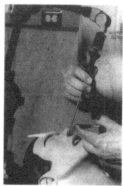 

Figure 13.9. VR simulators for medical applications. The biomechanical virtual reality model for the lower extremity (*a*) can be interactively evaluated (from Pieper *et al* 1991). The outcome of hip joint replacement is derived from the position of the tendons. A physical model with an endoscope and forceps (*b*) simulate endoscopic sinus surgery (from Edmond *et al* 1998). A laparoscopic simulator with force feedback (*c*) is used for the simulation of liver surgery (from Marescaux *et al* 1998).

Some commercial products for the simulation of laparoscopic procedures, ureteroscopy, neuroendoscopy and minimally invasive surgery are described by Chen and Marcus (1998). Simulators for central venous catheter placement, for individual tasks such as ligating and dividing, catheter-based endovascular therapy, balloon angioplasty and stent placement are mentioned elsewhere (Satava and Jones 1998). Many other surgical simulators are currently being developed at research centers and companies around the world.

13.4.2 Simulating on patient-specific data

Besides modeling on generic data sets such as the Visible Human, patient-specific data sets can be used for VR simulations. Virtual environments can be used to predict the outcome of surgeries on individual patients. Different procedures and treatments can be tested in the VE and the optimal startegy for each patient can be determined. For example, simulators may allow the surgeon to plan and simulate the outcome of procedures for plastic and reconstructive surgery (Rosen 1998).

13.4.3 Tissue modeling

Most of the VR applications in medicine do not limit interaction to simple visual inspection. Different from flight simulators, where contact with the environment is excluded or will end the simulation, interaction with the objects is a necessity in medical simulations. The environment is explored and changed by cutting tissues, feeling resistance, etc. Organs can be flexed, cut, tugged, stretched and reattached. Physical models of the human body, instruments and the operating room are required to simulate surgeries.

Possible approaches for the modeling of tissue behavior include surface or volumetric tissue models, spring models and finite element models (Delingette 1998). Unfortunately most of these models require long computation times. However, realistic tissue modeling for medical simulation has to be accurate and fast (20–30 frames per second). Achieving both goals is not possible with today's computers and thus they have to be compromised depending on the application.

Finite element meshes (FEM) can be used to model deformation of tissues (Bro-Nielsen 1998). The FEM divides complex structures such as the human body into many elements. The elements are very small and it is assumed that they have homogeneous tissue properties (e.g. elasticity) across each individual element and share boundary conditions with their neighboring elements. The outcome of a FEM simulation can be calculated by various mathematical algorithms.

13.5 IMAGE GUIDANCE

Virtual and augmented reality systems are not limited to diagnosis, planning and simulation but can also be beneficial during surgery. Chapter 12 discusses the use of different types of image guidance in operating rooms. Several sophisticated systems allow for the fused visualization of pre- or intraoperative images and outlined structures such as tumors or other critical areas in real time. One example is shown is figure 13.10 from the Surgical Planning Lab at Brigham and Women's Hospital, Harvard Medical School.

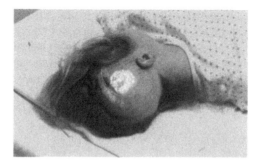

Figure 13.10. Intraoperative image guidance for neurosurgery at the Brigham and Women's Hospital. Live video images are registered and superimposed with preoperative data. A brain tumor is outlined to the surgeon (from Grimson *et al* 1999).

Virtual reality sensors and actuators can further enhance these systems, for example by providing the surgeon with 'eyes' that can see more and with 'hands' that can feel differently or operate with scaled forces and precision. Potential benefits include increased patient safety, higher precision, faster and more cost-effective procedures, and innovative surgery techniques. The challenges for these systems include the real-time aspect and the registration of objects that may deform during interventions.

Levy *et al* (1998) report the design of a special HMD for the use in microsurgical and endoscopic minimally invasive procedures. The HMD provides stereoscopic vision to replace the standard 2D images on a monitor. The surgeon can also switch the display to computer-generated images instead and thereby access additional information. Another advantage of the system is that the surgeon has an optimal view independent of the orientation of his hands and head. This is not necessarily true when a monitor is used. Identical images can be displayed in a HMD of surgery assistants so that they can better support the procedure. The HMD was reported to remain comfortable over several hours in real surgeries. One of the current limitations is the bulk and weight of the device and the limited spatial resolution (640×480).

Wagner *et al* (1997) used HMDs to partly immerse the surgeon in an augmented reality where the real world was captured with live video images and registered with presurgical image data (video see-through HMD). This approach provides visual access to invisible data of anatomy, physiology and function to improve osteotomies of the facial skeleton. The real-time registration was achieved with electromagnetic tracking of the subject's anatomy, medical instruments and the imaging devices.

The registration of ultrasound images with live video images for breast biopsy is reported by State *et al* (1996). Their research was carried out at the University of North Carolina, Chapel Hill, to simplify both learning and performing ultrasound-guided interventions.

13.6 TELESURGERY

Telemedicine applications are already extensively used in some disciplines of clinical practice such as radiology. The physician can diagnose the images after their acquisition in a different division of the hospital, at home, or even in a different city by electronically transferring the images to the desired location. There are also applications where physicians can diagnose patients with video-conference links from a remote site. Such set-ups may increase the cost effectiveness but concerns about impersonal care may rise (Hutchinson 1998).

A logical extension of the concept of remote diagnosis is the idea of surgery at remote, inaccessible or hazardous locations. Applications for such telepresence technology do not only include the operation of patients on distant sites such as a space station, submarine, plane, boat, or battle field. They can also be used in minimally invasive surgery to overcome problems in the mechanical control of conventional devices by restoring the hand–eye coordination that is lost otherwise. In microsurgery, the motion, forces and visual feedback can be scaled to the appropriate dimensions and thereby allow new procedures (Hill and Jensen 1998). Another potential benefit could be to limit technology and advanced therapy to few centers for quality control and better cost effectiveness.

Telepresence systems require a virtual environment that provides the surgeon with sufficient capabilities and feedback to perform challenging tasks successfully. This includes tactile feedback, stereoscopic and stereophonic data acquisition and display, and high-speed communication links for control of the surgical tools in real time. Figure 13.11 shows a model for a telesurgery and telepresence system.

Telesurgery has many potential applications, but there are many obstacles to overcome. Most importantly, the current systems need improved force and tactile feedback and faster data links to minimize the lag time for more realistic scenarios. In addition, optimal systems must be able to perform many complex tasks such as cutting, suturing and dissecting with intuitive and easy user interfaces. Research in this field is continuing and some systems for fields such as ophthalmic and inner ear surgery exist (Hill and Jensen 1998).

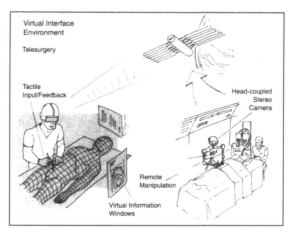

Figure 13.11. A model for telesurgery. The physician and the patient at a remote location are coupled with a telecommunication link. The physician receives stereoscopic, stereophonic and tactile impressions from sensors at the surgical site. The surgery is performed by the physician(s) on a virtual model and by a surgical robot on the patient (from Rosen *et al* 1996).

PROBLEMS

13.1   List and describe the sensors and actuators of humans.
13.2   List and describe sensors and actuators of virtual environments.
13.3   Explain how a VR data glove measures hand movements.
13.4   Explain the principle of stereoscopic image display.
13.5   Compare augmented and virtual reality.
13.6   List applications of VR for minimally invasive medicine.
13.7   List applications and advantages of virtual simulators.
13.8   Explain the challenges for virtual simulators.

REFERENCES

Ackerman M J 1998 The visible human project *Proc. IEEE* **86** 504–11
Bro-Nielsen M 1998 Finite element modeling in surgery simulation *Proc. IEEE* **86** 490–503
Burdea G C 1996 *Force and Touch Feedback for Virtual Reality* (New York: John Wiley)
Chen E and Marcus B 1998 Force feedback for surgical simulation *Proc. IEEE* **86** 524–30
Delingette H 1998 Toward realistic soft-tissue modeling in medical simulation *Proc. IEEE* **86** 512–23

Delp S P and Zajac F E 1993 Force- and moment-generating capacity of the lower extremity muscles before and after tendon lengthening *J. Biomech.* **26** 485–99

Dumay A C M 1996 Beyond medicine *Eng. Med. Biol.* **15** (2) 34–40

Edmond C V Jr, Wiet G J and Bolger B 1998 Virtual environments—surgical simulation in otolaryngology *Otolaryngol. Clin. N. Am.* **31** (2) 369–81

Faulkner G and Krauss M 1996 Guidelines for establishing a virtual reality lab *Eng. Med. Biol.* **15** (2) 86–93

Frankenthaler R P, Moharir V, Kikinis R, van Kipshagen P, Jolesz F, Umans C and Fried M P 1998 Virtual otoscopy *Otolaryngol. Clin. N. Am.* **31** 383–92

Gorman P J, Meier A H and Krummel T M 1999 Simulation and virtual reality in surgical education: real or unreal? *Arch. Surg.* **134** 1203–8, webpage http://archsurg.ama-assn.org/

Greenleaf W J 1996 Developing the tools for practical VR applications *Eng. Med. Biol.* **15** (2) 23–30

Grimson W E L, Gering D T, Nabavi A, Kikinis R, Hata N, Everett P, Jolesz F, and Wells W M 1999 An integrated visualization system for surgical planning and guidance using image fusion and interventional imaging *MICCAI'99 Med. Image Comput. Comput.-Assisted Intervention — Second Int. Conf. Proc.* (Berlin: Springer) pp 809–19

Haluck et al 2000 Joint research project in surgical simulation, Department of Minimally Invasive Surgery, Penn State University College of Medicine, Milton S. Hershey Medical Center, Hershey, PA, webpage http://cs.millersv.edu/haptics/

Hill J W and Jensen J F 1998 Telepresence technology in medicine: principles and applications *Proc. IEEE* **86** 569–80

Hoffman H and Vu D 1997 Virtual reality: teaching tool of the twenty-first century *Acad. Med.* **72** 1076–81

HT Medical Systems 2000 webpage http://www.ht.com/

Hutchinson J R B 1998 Telemedicine in otolaryngology *Otolaryngol. Clin. N. Am.* **31** (2) 319–29

Jones S B, Kuppersmith R B and Satava R M 1998 Otolaryngology in the information age— enabling technologies for the future of surgery *Otolaryngol. Clin. N. Am.* **31** (2) 241–53

Levy M L, Chen J C T, Moffitt K, Corber Z and McComb J G 1998 Stereoscopic head-mounted display incorporated into microsurgical procedures: technical note *Neurosurg.* **43** 392–6

Marescaux J, Clement J M, Tassetti V, Koehl C, Cotin S, Russier Y, Mutter D, Delingette H and Ayache N 1998 Virtual reality applied to hepatic surgery simulation: the next revolution *Ann. Surg.* **228** 627–34

Massachusetts Institute of Technology 2000 webpage http://touchlab.mit.edu/

Neri E, Boraschi P, Braccini G, Caramella D, Perri G and Bartolozzi C 1999 MR virtual endoscopy of the pancreaticobiliary tract *Magnetic Resonance Imag.* **17** (1) 59–67

Pieper S, Delp S, Rosen J *et al* 1991 A virtual environment system for simulation of leg surgery *Proc. SPIE* **1457** 188–97

Robb R A 1996 VR assisted surgery planning *Eng. Med. Biol.* **15** (2) 60–69

Rosen J M 1998 Advanced surgical technologies for plastic and reconstructive surgery *Otolaryngol. Clin. N. Am.* **31** (2) 357–69

Rosen J M, Soltanian H, Redett R J and Laub D R 1996 Evolution of virtual reality *Eng. Med. Biol.* **15** (2) 16–22

Salisbury J K and Srinivasan M A 1997 Phantom-based haptic interaction with virtual objects *IEEE Comput. Graphics Appl.* **17** (5) 6–10

Satava R 1994 Medicine 2001: the king is dead, webpage http://www.csun.edu/cod/94virt/med~1.html

Satava R M and Jones S B 1998 Current and future applications of virtual reality for medicine *Proc. IEEE* **86** 484–9

Serra L, Kockro R A, Gim-Guan C, Hern N, Lee E, Lee Y H, Nowinski W L 1998 Multimodal volume-based tumor neurosurgery planning in the virtual workbench *MICCAI'98 Med. Image Comput. Comput.-Assisted Intervention—First Int. Conf. Proc.* (Berlin: Springer) pp 1007–16

Sharma R, Pavlovic V I and Huang T S 1998 Toward multimodal human–computer interface *Proc. IEEE* **86** 853–69

Soferman Z, Blythe D and John N W 1998 Advanced graphics behind medical virtual reality: evolution of algorithms, hardware, and software interfaces *Proc. IEEE* **86** 531–54

Sorid D and Moore S K 2000 The virtual surgeon *IEEE Spectrum* **37** (7) 26–31

Srinivasan M A and Basdogan C 1997 Haptics in virtual environments: taxonomy, research status, and challenges *Comput. Graphics* **21** (4) 393–404

State A, Livingston M A, Hirota G, Garret W F, Witton M C, Fuchs H and Pissano E 1996 Technologies for augmented-reality systems: realizing ultrasound-guided needle biopsies *ACM SIGGRAPH—Assoc. Comput. Machinery Special Interest Group Comput. Graphics Interactive Tech.* 439–46

Tang S L, Kwoh C H, Teo M Y, Sing N W and Ling K V 1998 Augmented reality systems for medical applications *Eng. Med. Biol.* **17** (3) 49–58

Teneo Computing 2000 webpage http://www.teneocomp.com/

The Virtual Reality in Medicine Lab 1999 University of Illinois–Chicago, Chicago, IL, webpage http://www.sbhis.uic.edu/VRML/*Publications/publications VTB01.htm*

Wagner A, Rasse M, Millesi W and Ewers R 1997 Virtual reality for orthognathic surgery: the augmented reality environment concept *J. Oral Maxillofacial Surg.* **55** 456–62

CHAPTER 14

MINIMALLY INVASIVE SURGICAL ROBOTICS

Glenn Walker

Robots have many advantages in the medical field. They do not get tired, they are capable of precise and steady positioning of tools, and over the long term they can be cost effective. Some disadvantages include electric interference to and from other systems and loss of control by the surgeon.

This chapter presents an overview of how different types of robot are applied to various aspects of minimally invasive surgery and, where applicable, gives results of trials on humans or animals.

14.1 INTRODUCTION TO ROBOTICS

Webster's New World Dictionary defines a robot as, 'A mechanical device operating in a seemingly human way'. The Robot Institute of America defines a robot as, 'A reprogrammable and multifunctional manipulator, devised for the transport of materials, parts, tools, or specialized systems, with varied and programmed movements, with the aim of carrying out specified tasks'. When asked to define what a robot was, Joseph Engleberger, who developed the world's first industrial robot in 1961, said, 'I know one when I see one!' There are as many interpretations of what a robot is as there are different types. In this section we discuss some of the terminology and aspects common to all robotic systems.

One common connection between all robots is versatility. Robots are designed to be used in a variety of environments with common objectives. An example in minimally invasive surgery would be a robot used for treatment of a brain tumor. The robot has to accomplish the same objective (treatment of the tumor) in many patients but may have to do it by using different paths for each because of anatomical variations.

14.1.1 Components of a robotic system

Robotic systems can be divided into five subsystems (Coiffet and Chirouze 1983). These are:

1. The *articulated mechanical system* (AMS). This is the structural part of the robot that moves to accomplish tasks. Items in the AMS include limbs, joints and end effectors (tools or hands at the working end of the robot).
2. The *actuators*, which are used to move the various components of the AMS. A common actuator is the electric motor, which is used to turn belts or gears that in turn move various robotic parts.
3. The *transmission devices or systems*, which are used to link the actuators and the AMS. Examples of these are belts, pulleys, gears, or any other linkage that can cause movement in components of the AMS.
4. The *sensors*, which are used to gather input for the robot about its current position and the conditions of its working environment. Sensors can also be used to receive and relay information to a human controlling the robot.
5. The *robot brain*, which is the computing hardware of the robot. The brain is used to process the sensor data and make decisions about future movements, in the case of autonomous robots. In human-controlled robots, this is simply the computer software/hardware used to coordinate various functions of the robot (e.g. movement).

Figure 14.1 shows a diagram of how all the subsystems of a robotic system fit together. A two-way link is shown between the robot and the brain so that the sensors can talk to the brain as well as the brain giving instructions to the robot. The actuator is an electric motor driving a pulley, which in turn moves the arm. Sensors are located on the robot at various locations (only two are shown for clarity). These sensors give information to the brain about the current position of the robot as well as the status of its limbs and end effector (e.g. how much force the arm is applying to some object, or at how much of an angle some arm is currently rotated). User input can be involved in semi-autonomous robots, or if an autonomous robot behaves incorrectly and requires human intervention.

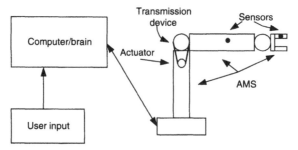

Figure 14.1. An example of the five subsystems of robots. The two-way link between robot and brain is used by the computer to control the robot but also by the sensors to communicate to the computer.

14.1.2 Conceptual models of robots

When discussing robots, a commonly encountered quantity is the number of degrees of freedom (DOFs) it has. Degrees of freedom can be explained as the number of independent movements an object can make with respect to a coordinate set R (Coiffet and Chirouze 1983).

One way to visualize a DOF is to imagine a rigid rod of some length in three-dimensional (3D) space. This rod can be translated along any of the three coordinate axes and it can also be rotated about any of the coordinate axes. Therefore, there are a total of six DOFs—three for translation and three for rotation. Figure 14.2 shows the concept of translation and rotation. It explicitly shows only two of the six total DOFs.

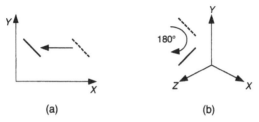

(a)   (b)

Figure 14.2. Two of the six degrees of freedom. (*a*) The rod is translated along the *x*-axis. (*b*) The rod has been rotated 180° around the *z*-axis.

By using these six DOFs, a robot can position its end effector (tool) at a given point and in the correct orientation for all cases. A robot never uses more than six independent DOFs. Even if it has many joints and limbs (i.e. degrees of mobility), their combined efforts are largely redundant and do not result in more than six independent DOFs.

Most robots used in surgical applications have six DOFs. Building a robot with this many DOFs allows the robot to be versatile and thus able to handle unexpected positioning complications that might arise during an operation. Six DOFs are not always needed for simple tasks that never present unexpected obstacles, like loading cement bags onto a truck.

14.1.3 Robotic control

Robots are generally composed of several movable components (limbs, effectors, etc). All of these must be controlled by the brain system of the robot. The purpose of robotic control can be stated succinctly as trying to match the coordinate vector of the end effector of the robot, X, with the coordinate vector of the point of interest P (e.g. tumor in the brain or polyp in the colon). Very advanced robots, which are still in development, can be given commands and can perform autonomously to achieve that command. Most robots require at least some human intervention.

In order to successfully control a robot, a model of the AMS and end effector must be created by which the brain can calculate positional changes. The computer calculates positional changes using a mathematical model of the robot. It then implements these changes by moving the joints and end effector of the robot. There are two types of model: geometrical and kinematic.

14.1.3.1 Geometrical model. A geometrical model of a robot treats each limb as an infinitely rigid member joined by perfect joints (e.g. no friction). Also assumed is that each member can have only one DOF with respect to either limb to which it is attached. Each of these ideal segments has its own coordinate system associated with it. Therefore, each segment moves relative to the preceding segment's coordinate system in only one DOF (i.e. either a translation or rotation). The coordinate system of the segment of the robot that is fixed (attached to something stationary) is called the task coordinate set.

The position of the end effector, with respect to the task coordinate set, is a function of the difference between the initial values (i.e. angles and translations) of the joints and the current values:

$$X(R_0) = F(\Theta - \Theta_0) \qquad (14.1)$$

where X is the vector representing the position of the end effector, R_0 is the task coordinate set, and Θ is the vector representing the current positions (angles and amount of translation) of all the joints.

In most applications of robotics, the position is known and the values of the joints need to be found. These values can be calculated by inverting equation (14.1):

$$\Theta - \Theta_0 = F^{-1}(X(R_0)). \qquad (14.2)$$

If equation (14.2) can be solved then that configuration for the robot is said to be resolvable.

In some cases there can be many possible configurations for the final Θ vector. Some of these positions may not be acceptable, especially for medical robots, as they might cause harm to the patient or observers. Also, some solutions to equation (14.2) may be unrealizable physically with the robot. The assumptions of rigidity and no friction can also cause positioning problems.

14.1.3.2 Kinematic model. The kinematic model is a linear approximation to the geometric model. Linearization is made possible by treating the continuous movements in the geometric model as a sequence of small incremental movements in the kinematic model. Equation (14.1) can be rewritten kinematically as

$$\Delta X = \frac{\partial F}{\partial \Theta} \Delta \Theta . \qquad (14.3)$$

As with the geometrical model, the kinematic model also assumes ideal structural members and joints, which can lead to inaccuracies in positioning. However, the kinematic model offers a greater control over the velocity of the limbs since the incremental movements, $\Delta\Theta$, which can be adjusted in speed, take place over a known time.

14.1.4 Robotic actuators

Actuators are the devices used to move the limbs of robots. They can be mechanical (e.g. hydraulic or pressure) or electric (e.g. electric motors). Actuators are controlled by the brain of the robot or by a human operator. They are designed and implemented in the robot from the standpoint of servo-systems, or servo-control theory. A servo-system is a system that takes an electric input and implements it as an electric or mechanical output. In the case of robotics, the output is usually a change in position of a limb or end effector. A detailed discussion of servo-systems is beyond the scope of this chapter, but the interested reader is directed to Coiffet and Chirouze (1983), or others devoted to servo-system theory.

14.1.5 Robotic sensors

Robots are equipped with numerous sensors that give them information about their position with respect to their base (internal sensors) as well as to environmental landmarks (external sensors). Internal sensors include potentiometers, variable transformers, optical encoders and strain gages. Potentiometers can be used by the robot to calculate how much one of its limbs has translated or rotated. Variable transformers are an alternative to potentiometers and measure the same movements. Optical encoders are used to provide information about how much a joint has rotated. The robot can also use these three sensors to deduce the speed at which a limb has moved. The stress in a limb can be used to calculate how much pressure is being applied to an object by the robot. Strain gages are used for this purpose and also for calculating the acceleration of a limb.

External sensors used by the robot include tactile sensors, proximity sensors and visual sensors. Tactile sensors are used to determine the amount of force the robot applies to an object. These types of sensor usually consist of analog circuits that produce a voltage proportional to the amount of stress on the sensor. Proximity sensors are used to detect objects in the robot's workspace. The main reasons for these are safety to humans or other objects in danger of being harmed by the robot, and also the ability of the robot to detect an imminent collision and avoid it. Ultrasound and infrared are two common modalities used for proximity sensors. Visual sensors, the most common of which are charge-coupled devices (CCDs), give the robot a crude ability to see various objects. Visual sensors greatly enhance a robot's ability since, among other things, they allow for recognition of objects.

14.2 MEDICAL ROBOTICS

In the future, the use of robotics to augment surgeons' ability will provide better results at lower cost. Taylor (1999) gives five areas in which traditional surgical techniques have reached a point of diminishing returns and can be augmented by robotics: planning and feedback, precision, access and dexterity, accuracy, and specialization and access to care.

Planning and feedback are used by the surgeon to locate the diseased tissue. A surgeon normally looks at various images of the patient. He or she then uses this information to go to the site and treat the disease. However, it is difficult for a surgeon to use pictures taken from one machine and to keep them in mind as they cut into a patient, in search of the diseased tissue. Robots could much more easily incorporate the different images into one picture and use that picture during their motion-planning phase.

Precision is the ability to suture small blood vessels or to perform surgery on delicate and/or small organs (e.g. the eyes or the brain). There is a limit to the spatial resolution of the human hand. In some surgeries, even the smallest slip or tremor by the surgeon could cause irreparable damage. Inconsistencies in the force used by the surgeon to cut or move tissue could also cause damage. Robots can overcome this problem as their precision can be made as small as technology currently allows. Also, they can be programmed to provide continuous movement with a continuous force and/or speed.

Access is how the surgical site is accessed and dexterity is the ability of the surgeon to move around in the area to complete the procedure. Minimally invasive procedures use small entry points for access, which greatly limit the surgeons' dexterity. However, the main advantage is that by using smaller entry points, the trauma to the patient is greatly reduced, thus reducing the chance of complications, the length of hospital stays, and ultimately cost. Robotics can give the surgeon the dexterity he or she needs while still using a minimally invasive approach. An example of this is the robotic endoscopes and laparoscopes, discussed later in the chapter.

Accuracy is the ability of the surgeon to cut straight lines or perform other tasks that require geometrical consistency. One example is in orthopedic surgery where accurate machining of the bone is needed so that the prosthetic will fit. Robots are ideal for these types of situation as they can be programmed to carry out tasks consistently and to very high degrees of accuracy.

Specialization and access to care is an important issue in health care since in most parts of the world there is a shortage of doctors, and especially specialists, in rural areas. Robotics, along with telemedicine, will all but eliminate this problem. By having a specialist in any part of the world, he or she could sit down at a terminal that is linked to a robot on the other side of the world. They could then perform surgery as if they were in the operating room. Such devices will have a profound impact, especially in developing countries.

The following sections discuss how robotics has been applied to various surgical procedures. Some of the robots mentioned have already been used many times in real surgeries.

14.2.1 Robotic endoscopes

An endoscope is a device that allows surgeons to access and see a site through a small hole in the body. The endoscope can also be fitted with tools to perform surgery (e.g. removal of polyps with a lasso). Unfortunately, the minimally invasive advantage of the endoscope is also its disadvantage—it limits the surgeon's ability to maneuver in and view the surgical area.

To alleviate this problem, researchers have attempted to make certain aspects of the procedures more effective via robotics. One example is repositioning tools within the patient. A normal endoscope has channels along its axis that can contain tools inserted from the outside. The position (up/down and left/right) of an endoscope within the patient is controlled by changing the external controls that in turn bend the last 10 cm of the endoscope to the desired position. In order to position the tool in the proper location, the entire last 10 cm must be moved. This coarse movement of the endoscope can complicate procedures by deforming the work area. It also might prove too clumsy for delicate procedures. A solution to this problem is to place a movable platform at the end of the endoscope that can direct tools in a more precise manner.

One device that accomplishes this is the endo-platform (Cohn *et al* 1995). The endo-platform provides fine control of tools at the very end of the endoscope separate from the coarse controls used to move the entire scope. Figure 14.3 shows that two plates separated by a spring make up the endo-platform. Tendons attached to the outer plate can be adjusted by the surgeon to move it with respect to the inner plate. A joystick controller allows the positioning of the tool in any hemispherical area relative to the current endoscope position.

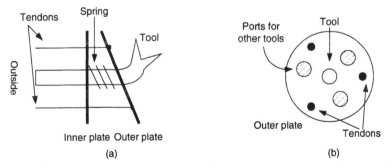

Figure 14.3. The endo-platform. (*a*) Side view. The distance between the plates is 20 mm. (*b*) Top view. The outer plate is moved relative to the inner plate via the tendons. In this way the tool can be repositioned within a patient. Ports cut into the endoscope allow for other tools to be inserted simultaneously.

A prototype endo-platform has been built in which the diameter of the plates is 19 mm with a distance of 20 mm between the plates. Three tendons are attached to pulleys on dc servo-motors, which control the positioning of the outer plate. A closed-loop controller with an update rate of 500 Hz is used to position the end plate.

14.2.2 Gastrointestinal endoscopy

Another way in which robotics have been applied to endoscopy is through the development of a device that examines the small intestines. The small intestines are far enough up the gastrointestinal tract that they cannot be reached by conventional endoscopes. To solve this problem, a robot has been developed that can crawl up into the small intestines without pushing from the outside by the surgeon (Slatkin *et al* 1995).

Figure 14.4 shows an overview of the system. The robotic endoscope is controlled via a computer. It consists of a 1.4 mm diameter endoscope in which fiber optics are used for illumination and imaging within the small intestines. Electric wiring from the computer is used to control the movement of the robot. Tubing is used to connect high- and low-pressure sources to the grippers and extensors.

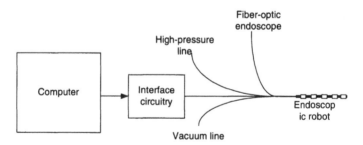

Figure 14.4. A computer is used to control the endoscopic robot. High-pressure and vacuum lines are used for the extensors and grippers. The fiber optic is used as a camera.

The movement mechanism of the robot is designed specifically for the small intestines, which are very convoluted, slippery, and can change in diameter up to fourfold. The movement cycle, or gait, is analogous to the movement of an inchworm. The robot uses grippers on either end to provide traction by attaching to the current position in the small intestines. The grippers, when contracted, are 22 mm in diameter and 38 mm in length. They are balloons that can be inflated and deflated via the external pressure lines.

The extensor section, which is in between the grippers, can increase in length or decrease in length. When fully contracted, it is 34 mm long. It can extend to approximately 51 mm. Extension and contraction is accomplished by pumping gas in and out of modified bellows. Figure 14.5 shows the gait of a two-gripper, one-extensor, section of the robotic endoscope (figure 14.4 shows a five-section robot). Any number of these segments can be combined to create a longer robot.

If a longer robot is used, several gait patterns can be established. The simplest would be to cycle through each of these six positions for each segment, until each segment in the robot has been moved. The cycle would then start all over again.

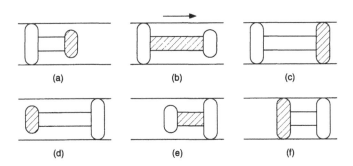

Figure 14.5. Step-by-step gait of the robotic endoscope. (*a*) The front gripper deflates. (*b*) The extensor extends, using the rear gripper to hold on to the intestines. (*c*) The front gripper inflates. (*d*) The rear gripper deflates. (*e*) The extensor shrinks and pulls the rear gripper towards the front. (*f*) The rear gripper inflates and the process starts over again with (*a*).

The computer controls the action of the grippers and extensor to provide movement. The user simply specifies direction and speed with an external control unit.

Slatkin *et al* (1995) conducted preliminary studies using one segment of the robot in a section of swine intestine. The robot was able to move a little, but further development is needed before more thorough animal studies can be conducted.

14.2.3 Colonoscopy

Colonoscopy is traditionally performed with an endoscope in the colon. It is considered a minimally invasive technique since no incisions need to be made and one device, the endoscope, can accomplish diagnosis via pictures, and therapy via actuators attached to it. One main drawback of the use of endoscopes in examining and treating the colon is discomfort to the patient. The endoscope is fairly rigid with only the front end being flexible. Applying robotics to this type of examination would cause less discomfort through more flexible instruments designed to navigate the convoluted environment of the colon.

Figure 14.6 shows an endoscope developed by Carrozza *et al* (1997) for navigating the colon. It is composed of a one-segment robot that can navigate the colon, tools for diagnosis and intervention, and a user interface on a computer.

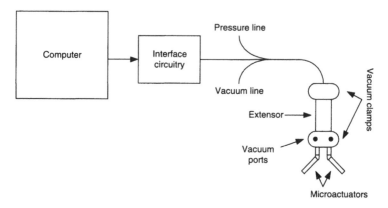

Figure 14.6. As in figure 14.4, a computer is used to control the robotic colonoscopy system. Vacuum and pressure lines are used for the extensor and grippers. The microactuators are conceptual and were not tested in the prototype.

The robot moves by inchworm motion, as in the endoscopic robot reported by Slatkin *et al* (1995). However, the grippers on either end of the colonoscopic robot use a vacuum instead of a balloon as its attachment mechanism. The colonoscopic robot is also only one segment long. The dimensions of it are 50 mm when contracted and 80 mm when fully extended. It has a diameter of 18 mm. Plastic tubes run through the robot to the outside of the patient to allow for external insertion of devices (e.g. fiber-optic cameras or cauterizers).

This device was tested *in vitro* with a restructured swine colon to provide a more human-like model. The investigators found that the robot was able to successfully maneuver in the swine colon both forwards and backwards in a consistent manner. Further tests are planned.

14.2.4 Laparoscopy

Laparoscopy is the use of an endoscope and surgical tools to access the abdominal cavity, usually through one or more small incisions on the outer surface of the skin. Unlike other endoscopic procedures, the tools are usually inserted through openings separate from the camera. Because of this, positioning of the camera is more difficult, and more critical, than in other types of endoscopic procedure.

Positioning the camera is normally done with a surgical assistant who uses his/her judgement and follows directions from the surgeon. One application of robotics is to create a mechanized arm that acts as a camera holder and positioner, and is controlled by the surgeon. Advantages to this would be increased accuracy, steadiness of the images, and over the long term, the elimination of costly surgical assistants. One robotic camera positioning system currently on the market accomplishes this (AESOP, Computer Motion, Goleta, CA). AESOP is an acronym for automated endoscopic system for optimal positioning.

AESOP requires human direction about where to position the camera. At least one research group has made progress towards developing a laparoscopic camera that follows the surgical site automatically through computer vision analysis (Casals *et al* 1996). Figure 14.7 shows an AESOP system.

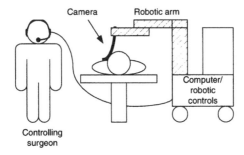

Figure 14.7. The AESOP system. A surgeon controls the position of the camera via voice or hand, while at the same time operating on the patient.

14.2.4.1 Case studies. Laparoscopy is one of the few medical procedures in which robots are well developed enough to have been tested on animal and human subjects. Gagner *et al* used a robotic arm to hold a camera, which provided a view of the operating region within the patient (Gagner *et al* 1994). The robotic arm (A460 CRS Plus, Burlington, Ontario, Canada) was modified to hold a laparoscopic camera. The robot was controlled from another room by an operator via a joystick and two buttons. These controls provided up/down and left/right movement as well as in/out and rotation control. The operator, using a video monitor of the surgical site, continuously adjusted the robot to provide the best image for the surgeon. This robotic camera holder was used to assist in laparoscopic cholecystectomy (excision of the gall bladder) for three patients. No intraoperative complications or technical problems were encountered.

Kavoussi *et al* (1995) performed a comparison of robotic versus human control of a camera during laparoscopic urological surgery in 11 patients. Eight of the procedures were diagnostic laparoscopic pelvic lymphadenectomies (surgical removal of a lymph node) and three were Burch bladder suspensions. The AESOP robot was controlled by a foot pedal and was used by the surgeon to position the camera. The camera was operated through an 11 mm port. The human-controlled camera was positioned by a surgical assistant who moved the camera on directions from the surgeon. A comparison was made by one individual reviewing videotapes from both cameras. It was found that the robot was able to control the camera more precisely and hold it steadier than the human. However, the robot-assisted procedures were not accomplished any faster than those with a human-held camera.

Partin *et al* (1995) reported on 17 laparoscopic procedures that were performed by one laparoscopic surgeon who used one or two robotic arms to assist. The procedures included four nephrectomies (removal of a kidney), two retroperitoneal lymph node samplings, two varix ligations, three pyeloplasties (reparative surgery on the pelvis of the kidney), two Burch bladder suspensions, one pelvic lymph node dissection, one orchiopexy (suturing of an undescended testicle in the scrotum), one ureterolysis (loosening of adhesions around the ureter), and one nephropexy (surgical attachment of a floating kidney). An AESOP robot was used for the procedures through a 10 mm laparoscopic port. The motion of the camera robot was controlled with foot pedal. The second robotic arm, when used, functioned as a retractor and was manipulated by a hand control. The authors found that it was feasible to use robotic arms as surgical assistants in genitourinary (with respect to the genitals and urinary organs) laparoscopic surgery. The main long-term advantage, they argue, is the cost effectiveness of robotic controllers versus human operators.

Mettler *et al* (1998) used the AESOP robot over a period of 1 year on 50 patients undergoing routine gynecological endoscopic surgical procedures. They found that the robotic arm allowed two doctors to do the procedure much more quickly. It was also found that the voice-controlled version of AESOP works more efficiently than the hand- or foot-controlled version.

Margossian *et al* (1998) used a robotic system to laparoscopically reanastomosize (reconnect) the uterine horns of six female pigs. The robotic device was used for microsuturing. The robotic system used (ZEUS, Computer Motion, Goleta, CA) is specifically designed for microsuturing procedures. It consists of three robotic arms—one for holding the camera and two for controlling surgical instruments. The camera can be controlled with a foot pedal or with speech. The surgical instruments are controlled with handles in front of the surgeon. The surgeon sits away from the operating table in a chair in front of a monitor that provides the laparoscopic camera view. This system has the advantages of reducing fatigue of the surgeon and the ability to achieve high precision movements. The ZEUS system also allows the surgeon to scale his/her movements. For instance, a 15 mm movement at the controllers could translate to a 1 mm movement at the robotic grippers in the surgery site. The surgery on the six pigs was complete with only minor technical problems (e.g. loosening of one of the surgical tool graspers on the robotic arm). Also, the surgeons reported only mild fatigue. Figure 14.8 shows the ZEUS system.

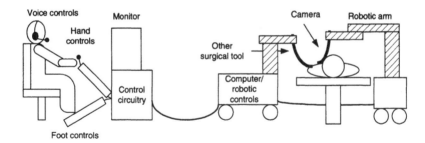

Figure 14.8. The ZEUS robotic system. The surgeon controls everything from a remote operating station. One of the robots holds a camera while the other holds a surgical tool to perform an operation.

14.2.5 Neurosurgery

New imaging technologies now allow doctors to locate regions within the brain that are of interest. Traditionally, in order to access these regions for surgery, the skull would have to be opened and a surgeon would manually perform some type of therapy without being able to see the site. The high degree of precision necessary for this type of procedure is obvious. Because of this, robotic systems are being developed that perform and assist in various neurosurgical procedures based on information acquired from different imaging technologies.

In the past, stereotactic frames have been used, which were physically attached to the patient's head. The patient was then scanned and the frame was visible in the scan. Trigonometric transformations could be used to determine how best to approach the area of interest from outside the skull. The surgeon could then position a probe guide, which was attached to the stereotactic frame, in the proper position and then perform surgery. The main disadvantage to this method was that repositioning the probe guide to access more than one site required much work and time, mainly because of the necessity to maintain a sterilized operating environment and of the high degree of accuracy needed in positioning the probe guide. One possible solution to this was to get rid of the probe guide completely and attach a robot, which could use the stereotactic frame as its reference frame, and could position a probe guide attached to its end effector automatically and quickly to the proper position.

The basic components of a neurosurgical robotic system are a controller, an interface, and the robot. The controller sits between the interface and the robot. It translates desired positions displayed on the interface into actual motor movements in the robot. The interface is used by the surgeon to observe the current state of the robot and where it is going. The surgeon also has the ability to shut down the robot or enter new commands. The robot simply follows the commands dictated by the interface and hence by the controller.

In neurosurgery, one of the main problems is how to most effectively get to the desired location. That is, the surgeon can identify an area of interest from brain scans (CT, MRI, etc) and then knows where the operation needs to be performed. The surgeon then identifies this point to the computer system. It is up to the computer system to calculate the most effective trajectory of the robot to reach this point.

Maciunas (1993) describes one robot that has been used for CT-directed stereotaxis: the Unimation PUMA 200 industrial robot (Westinghouse Electric, Pittsburgh, PA). The PUMA is a robot that consists of a single arm that is slightly larger than an adult arm, going from midtorso down to the hand. There are several characteristics of this robot that make it attractive for stereotactic neurosurgery (Kwoh *et al* 1988). It is programmable and versatile, it is designed for accurate, delicate work, it has a relative accuracy of 0.05 mm, and it is safe, as some of its joints are equipped with brakes that automatically clamp should any problems arise.

The PUMA is a six DOF device connected by a series of links. As mentioned in section 14.1, each link has its own coordinate system, with the first link being referenced to the real world. The reference frame of each link can be transformed to the next link with the matrix T_x:

$$T_x = \begin{bmatrix} \cos\theta_x & \sin\theta_x & 0 & 0 \\ -\sin\theta_x & \cos\theta_x & 0 & 0 \\ 0 & 0 & 1 & L \\ 0 & 0 & 0 & 1 \end{bmatrix}. \tag{14.4}$$

The 3×3 matrix in the upper-left part of equation (14.4) represents the rotation transformation from link to link, while the 3×1 matrix from the upper-right down represents the translation transformation. Since there are six links, the coordinates of the end effector, or in this case the probe guide, can be translated into the base (reference) coordinates, or vice versa, with the matrix product D:

$$D = T_1 T_2 T_3 T_4 T_5 T_6. \tag{14.5}$$

Each joint between each link has an angle, θ_x, associated with it. These form a vector

$$A = [\theta_1, \theta_2, \theta_3, \theta_4, \theta_5, \theta_6]. \tag{14.6}$$

The problem of positioning the robotic arm then becomes one of finding a set of joint angles that will position the probe guide at some point d. Or, by equation (14.2),

$$A = f^{-1}(d). \tag{14.7}$$

It is obvious that if the *d* is known, as in a surgeon who looks at a brain scan image and sees where the therapy needs to be performed, then there are infinitely many solutions to get to that *d*. What is more, all the *d* values that the computer comes up with may not be acceptable. Some path plans may place the end effector through vital regions in the brain on the way to the treatment area. To prevent this, training sessions are normally done with the robot before the procedure to eliminate unacceptable paths. The surgeon also reviews the plan before it is begun.

Once patients are scanned, they are removed from the scanner and the robot is attached to the base ring of the frame attached to their head. The computer system performs its motion planning and presents it to the surgeon. The surgeon then approves and watches as the robot positions the probe guide. Kwoh *et al* (1988) have tested their system on watermelons with small metal balls in them. They have also reported a successful biopsy on a 52-year-old patient. Figure 14.9 shows how a PUMA robotic arm is set up to position the probe guide over the patient's head.

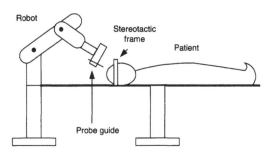

Figure 14.9. The PUMA robot is controlled by a computer and uses the information gained from the CT, or other types of imaging modalities, to find the optimal position of the probe guide.

Chen *et al* (1998) have used a PUMA robot to perform neurosurgery. Their system consisted of a computer system for image-guided planning and a PUMA 260 robot with a custom-designed end effector to hold surgical tools. Scans of the brain were acquired and then a 3D model was built. Surgical planning was done with custom-designed software, which allowed the performance of the robot to be evaluated. According to the authors, the robot has been applied successfully to the stereotactic neurosurgery clinical environment, although no details were given.

Drake *et al* report on using a PUMA 200 robot to hold a retractor during resection of thalamic astrocytomas in six children (Drake *et al* 1991). The retractor was a metal cylinder 7.5 cm long and 2 cm in diameter. It was introduced manually and then placed into the robot's grippers. Once placed in the grippers, the retractor's position could be accurately traced by the computer system attached to the robot. The robot acted as a rigid frame that could be placed in various locations with a high degree of accuracy—a positioning error maximum of 2 mm. The authors found that by using the robot, the presurgical plan could be more closely followed since the retractor could be positioned with a high degree of accuracy. Using the robot also allowed the position of the retractor to be precisely located on the images that had been made of the brain (i.e. CT and MRI scans) since during the surgery it is attached to the stereotactic frame used during imaging.

Instead of using nonmedical robots (e.g. the PUMA) placed in a medical setting, some six DOF robots have been developed specifically for neurosurgical procedures (Masutani *et al* 1995). The robot is mounted above a patient's head, which is in turn attached to a stereotactic frame. The brain surface and functional areas were scanned and the region of interest was selected by the surgeon. A Sparc 10 ZX SUN workstation took from 4 to 9 min to calculate a path for the robot. Once calculated, the needle was inserted by the robot. The surgeon could confirm the position of the robotic needle via the computer at any time during the procedure. Studies were conducted on watermelons with their seeds as the target. The error of navigation was determined to be 1–2 mm, which makes it acceptable for clinical use.

Goradia *et al* (1997) investigated the accuracy and safety of a robot-assisted evacuation of intracerebral hematomas (ICH) as compared to conventional methods, simulated *in vitro*. They found that using the robot resulted in slightly longer procedure time, but augmented the surgeon's dexterity and comfort.

Lavallée *et al* describe a system in routine clinical use since March 1989 (Lavallée *et al* 1992).

14.2.6 Eye surgery

Robotics have also been applied to various types of eye surgery. One particular application is the use of robotics for automatic placement of bursts of laser light (laser photocoagulation of the retina). Laser photocoagulation can be used to treat diabetic retinopathy (disorder of the retina due to diabetes) and retinal tears. The retinopathy is treated by denaturing areas of the retina around the periphery. However, as many as 3000 bursts of light may be used per retina for treatment. Retinal tears can be treated by mending the tears with the laser.

Barrett *et al* (1996) have developed a prototype system that can control the diameter and depth of the lesions via feedback from lesion reflectance and that can perform lesion placement using retinal vessels as landmarks to compensate for patient movement. Ferguson *et al* (1996) have built a prototype for use in a clinical setting. The motivation for using robotics in this particular application is because the surgery is time consuming and also because the placement of the lesions manually is determined by the dexterity of the ophthalmologist. The system developed is called the Computer-Aided Laser Optical System for Ophthalmic Surgery (CALOSOS) and it can create lesions in the retina in minimal time with little or no human intervention.

One important aspect of this type of surgery is tracking the patient's eye. Movement of the eye during laser exposure could have disastrous consequences. Because of this, CALOSOS has the ability to track the patient's eye. It can compensate for small movements by adjusting the laser. However, if large eye movements are detected, it will shut off the laser and reposition. Lesion creation can only begin again if it is initiated from the computer controller.

Retinal tracking is accomplished by using six (three horizontal and three vertical) variable width 1D templates using four pixels each of a CCD acquired image of the retina. Retinal blood vessels are used with the templates because of their sharp contrast with other features in the eye.

The six 1D templates are combined into a 2D template. This 2D template is constantly scanned over the retina image to make sure the correct positioning of the retina is maintained. If the response from the template is not as large as it should be, then a loss of lock signal is sent and the laser is shut off. In this event, the six templates must be scanned again and a new 2D template must be formed.

14.2.7 Orthopedic surgery

Robotic systems have been applied to orthopedics. The main advantage to using robots in orthopedics is their precise and repeatable movements. This is especially important since bone is machined by the robot. However, orthopedic procedures are not minimally invasive in general and the robots are used not for minimally invasive procedures, but because of their ability to perform precise, repetitive motions.

One of the more famous orthopedic robotic systems is ROBODOC. ROBODOC was designed for machining the femoral bone to match the shape of the prosthesis for hip surgery. Orthopedic robotic systems have been used with great success. We will not discuss these systems in detail as they are not minimally invasive.

14.2.8 Radiosurgery

Radiosurgery uses beams of radiation to ablate tissue, usually within the brain, that cannot be reached easily by other means. The beams come from a single radiation source that is aimed at the desired location from different angles, so as to minimize the amount of radiation deposited in healthy tissue. Accuracy of most radiosurgery systems approaches 1 mm.

A patient first has a stereotactic frame attached to his or her head. They are then imaged with CT or MRI. The location of the region of interest is noted by the surgeon, as well as critical features that are sensitive to radiation (e.g. the optic nerve). Paths for the radiation beam and intensity for each path are selected by a surgeon. The patient is then radiated.

One area in which robotics, or robotic motion planning, can be applied is in determining the most acceptable paths for radiation. This is similar to the motion planning used by the PUMA robots, except here there is no probe guide—just a beam of radiation. There are only certain angles at which the radiation can be administered since some features in the brain are very sensitive to it. Minimizing radiation dosage to critical neural structures is of paramount importance in planning the sweep path.

Schweikard *et al* (1995) report on using a six DOF GM-Fanuc 420 manipulator, which moves a linear accelerator that creates a photon beam, to sweep a radiation source in arbitrary paths with adaptable collimator beams. They developed a program that asks the surgeon for the number of arcs to sweep through, as well as the length of each. The program then optimizes, based on the number of arcs entered by the surgeon, how to distribute the doses over all the sweeps. Figure 14.10 shows how two beams can minimize dosage to healthy tissue while damaging diseased tissue.

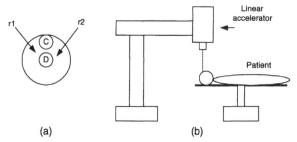

Figure 14.10. (*a*) Two beams (r1 and r2) from different angles affect healthy tissue slightly while damaging diseased tissue (D) and avoiding altogether critical tissue (C). (*b*) The robot holds the linear accelerator and is capable of moving it over the patient's head along various paths.

14.2.9 Ear surgery

Otosclerosis is a condition where the stapes becomes attached to the surrounding bone of the middle ear. This can lead to reduced hearing. However, this can be treated with surgery—namely a stapedotomy. In a stapedotomy, a prosthesis is attached from the incus directly to the fluid of the inner ear. This usually restores much of the patient's hearing. Figure 14.11 shows an ear without and with a prosthesis.

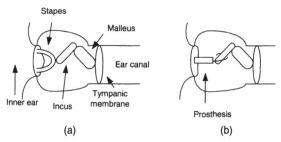

Figure 14.11. (*a*) A normal ear. (*b*) The stapes has been removed and a prosthesis inserted which connects the incus to the inner ear.

Drilling the hole through the stapes to get to the inner ear requires a great deal of skill, as the thickness is not known ahead of time and it is quite thin. It is quite hard for the surgeon to tell when he or she has broken through since there is very little perceivable change in force required to hold the drill.

Because of these conditions, robotics provide an ideal solution. They are able to provide repeatable motions, are very precise, can advance at very slow rates, and can have sensors mounted on them to detect the minute changes in pressure when the stapes has been successfully drilled through.

Baker *et al* (1996) have designed a system capable of drilling through flexible bone tissues during ear surgery. It is able to account for tissue deflections and thus detect when it is about to break through the tissue it is currently in. Figure 14.12 shows an overview of the system.

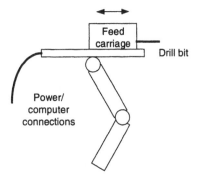

Figure 14.12. Stapedotomy tool. The bit is advanced into the ear canal and through the stapes.

Three sensors are used with the tool. A shaft encoder is used to monitor the position along the linear feed actuator. A full bridge strain gage is used to measure the force being applied by the drill bit. The torque is measured by monitoring the radial force on an idler gear.

Breakthrough can be signified when the force required to advance the drill begins to fall and the torque begins to rise. The tool is mounted on a support for stability as well as a constant feed rate. The support also allows the tool to be slid back quickly upon breakthrough. Feed rates of the order of 0.1–1 mm min^{-1} are used for the procedures.

14.3 ROBOTICS IN TELESURGERY

Section 13.6 discusses the concept of telesurgery and its application to minimally invasive surgery. Telesurgery involves an operator, usually separated by some appreciable distance from the physical surgery site, a way to transmit the operator's intentions, and a device at the actual surgical location that can act as an output for the remote user. The output may be as simple as a TV screen or as complex as a robot.

Also important in the telesurgery system is feedback to the operator at the remote site. Visual feedback is the most common and important. Also important, but less common are audio and haptic feedback. Some new systems are even attempting to incorporate cutaneous and olfactory sensations as forms of feedback (Hunter *et al* 1994).

Robots are a key component of telesurgery systems, as they are the next best thing to being at the actual surgical site. Robots, combined with a variety of sensors, can give the surgeon sight, touch and the ability to perform actions on the patient.

Most current surgical robotic systems are used with the operator standing next to them in the operating room. However, there is no reason why any robotic system cannot be used as a telesurgical system. The operator is already separated from the actual robotic manipulators by the distance of the electric cables running from the controller or computer to the robot's arms, even if they are in the same room. Extending the distance of these cables, as in telesurgery, presents no difficulties other than that the surgeon will no longer be in the same room as the patient.

Mitsuishi *et al* (1998) give an example of a telesurgical system. They constructed a telemicrosurgical system in which a blood vessel with a diameter of 1 mm was sutured by robot manipulators that were controlled by an operator 700 km away. The system is composed of the surgery site and the operation room. The surgery site contained the blood vessel, robot manipulators, a movable microscope, camera and microphone. These were all linked via the Internet to the operation room. The operation room is where the surgeon is located. The room contained a movable display, camera, speaker and robot manipulators. The movements by the surgeon of the manipulators were translated via real-time controllers into movements of the robot manipulators at the surgery site. Figure 14.13 shows an overview of the system.

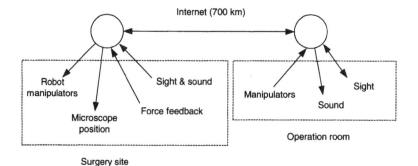

Figure 14.13 The surgeon is located in the operation room where he or she controls the robotic manipulators at the surgery site via the manipulators locally. The surgeon also receives visual and audio information from the surgery site. Visual information is recorded at the operation room and used by the controllers to position the display at the operation room and the microscope at the surgery site.

One unique feature about this system is its ability to adjust how the visual information recorded at the surgery site is presented at the operation room (Mitsuishi *et al* 1995). In order to accomplish this, a CCD camera is attached to the display at the operation room. The camera is positioned so that it constantly monitors the surgeon's face. If the surgeon turns his or her head left or right, the monitor adjusts itself so that it stays perpendicular to the surgeon's field of view. The CCD camera is also linked to the microscope at the operation room. If the camera detects that the user is looking more closely at the screen, it signals the microscope to zoom in.

An example of laparoscopic telesurgery is given by Schulam *et al* (1997). A surgeon experienced in endoscopic procedures was located 5.6 km away from the operating site, where a less experienced surgeon was performing the procedure. The remote surgeon was able to intervene if the less experienced surgeon ran into any difficulties.

A dedicated T1 line (1,544,000 bits per second), configured to carry voice or data traffic, was used to connect the two sites and provided enough bandwidth for video, audio and camera control. The remote surgeon had the ability to control the view of a camera in the operating room as well as switch to views from the endoscopic camera inside the patient. An AESOP robot was used to hold the endoscopic camera and could be controlled by the remote user.

Seven patients underwent procedures. There were no complications and it was the opinion of the remote surgeons that the system is a feasible solution for telesurgical consultation.

Graves *et al* (1997) are working on a system to provide mentoring of neurosurgical procedures in remote locations. The system is called the Integrated Remote Neurosurgical System (IRNS) and allows a remote neurosurgeon to control a robotic microscope in the operating room, communicate with others in the operating room through audio and video, and view presurgical images. The microscope is positioned by a robot in the operating room. As of the publication date, the system is still under development and no experiments had been performed.

Rovetta *et al* (1995) report on performing the first transatlantic experiment involving telesurgery. A robot in Milan, Italy, was used to perform surgery on a model of a pig while the controller was in Pasadena, CA. The two stations were linked by two communications satellites and a fiber-optic network. The robot was provided with a cutting tool, which it used to make an initial incision as directed by the operator in California. A surgeon in the operating room then inserted a trocar and fiber-optic camera into the pig. The robot then used a biopsy needle, which was advanced into the liver of the pig. The needle was removed and the material aspirated. The robot was then used to make an incision again through which the surgeon in the operating room could remove a cyst using forceps.

14.4 SAFETY

Safety is an important issue in surgery, especially when robots are involved. The concern people have about a nonhuman device performing potentially fatal actions on their organs is undoubtedly a major reason that robots have not become more commonplace in surgical theaters.

Some components of the robotic system that could be prone to failure are the brain (software), mechanical mechanisms and electric system. If there is a bug in the software, the results could be unpredictable and disastrous. An unexpected power failure or interference from other electronic equipment could lead to erratic motion of the limbs and end effector. A short circuit in the electric system, or interference, could lead to unpredictable outcomes as well.

A few precautions can be incorporated in the robot's design. One of these is to use only slow movements of the limbs. This would allow a possible fatal movement to be noticed before any damage occurs. Redundant sensors can be placed throughout the robot to make sure the limb positions are correct as observed by the robot. Mechanical constraints can be placed on the robot so that certain movements are not allowed.

Cosío *et al* (1997) discuss the susceptibility of robots to electromagnetic interference (EMI). There are two characteristics of robots that make them especially susceptible to EMI: different components of robots are often made by different companies, thus possibly having different grounding methods, and cables from the controls to the motors are sometimes quite long and shielding them is unfeasible.

Cosío *et al* (1997) discuss their experience with a robot used in an electrosurgical procedure for transurethral resection of the prostate (TURP). They found that the grounding provided with the robot was inadequate to deal with the EMI generated by the electrosurgical generator (ESG). They noted that the ESG, which is the unit used to generate electricity for electrosurgical applications, caused various errors with the robot. Errors included garbled communication of the sensors to the brain and malfunction of the motor controllers.

PROBLEMS

14.1   Define the word robot.
14.2   List the components of a robotic system and explain the purpose of each one.
14.3   List the six degrees of freedom. Think of a robot that has 20 joints. Explain whether or not this robot uses more than six independent degrees of freedom.
14.4   Describe the two types of robot model for control purposes.

14.5 List four areas in which robots can augment surgeons' abilities.

14.6 Explain why is it desirable to have a robotic endoscope as described in 14.2.2. Explain why a regular endoscope is not optimal.

14.7 Explain why it is critical that robots used for stereotactic neurosurgery be attached to the same frame from which the imaging is done.

14.8 Explain why it is advantageous to use robotic planning for radiosurgery.

14.9 Explain why it is advantageous to use a robotic drill for implanting a prosthetic stapes.

14.10 Describe some safety issues in dealing with robots.

REFERENCES

Baker D, Brett P N, Griffiths M V and Reyes L 1996 A mechatronic drilling tool for ear surgery: a case study of some design characteristics *Mechatronics* **6** (4) 461–77

Barrett S F, Wright C H G, Oberg E D, Rockwell B A, Cain C P, Jerath M R, Rylander H G III and Welch A J 1996 Integrated computer-aided retinal photocoagulation system *Proc. SPIE* **2673** 163–73

Borst C 2000 Operating on a beating heart *Sci. Am.* **283** (4) 58–63

Carrozza M C, Lencioni L, Magnani B, D'Attanasio S, Dario P, Pietrabissa A and Trivella M G 1997 The development of a microrobot system for colonoscopy *CVRMed MRCAS '97 First Joint Conf. Comput. Vision, Virtual Reality, Robotics in Med., Medical Robotics and Computer-Assisted Surgery Proc.* (Berlin: Springer) pp 779–88

Casals A, Amat J and Laporte E April 1996 Automatic guidance of an assistant robot in laparoscopic surgery *Proc. IEEE Int. Conf. Robotics and Automation (Minneapolis, MN, 1996)* vol **1**, pp 895–900

Chen M D, Wang T, Zhang Q X, Zhang Y and Tian Z M 1998 A robotics system for stereotactic neurosurgery and its clinical application *Proc. IEEE Int. Conf. Robotics and Automation (Leuven, 1998)* vol **2**, pp 995–1000

Cohn M B, Crawford L S, Wendlandt J M, and Sastry S S 1995 Surgical applications of milli-robots *J. Robotic Syst.* **12** (6) 401–16

Coiffet P and Chirouze M 1983 *An Introduction to Robot Technology* (London: Kogan Page)

Cosío F A, Hibberd R D, and Davies B L 1997 Electromagnetic compatibility aspects of active robotic systems for surgery: the robotic prostatectomy experience *Med. Biol. Eng. Comput.* **35** 436–40

Drake J M, Joy M, Goldenberg A and Kreindler D 1991 Computer- and robot-assisted resection of thalamic astrocytomas in children *Neurosurg.* **29** 27–31

Ferguson R D, Wright C H G, Rylander H G III, Welch A J and Barrett S F 1996 Hybrid tracking and control system for computer-aided retinal surgery *Proc. SPIE* **2673** 32–41

Gagner M, Begin E, Hurteau R and Pomp A 1994 Robotic interactive laparoscopic cholecystectomy *Lancet* **343** 596–7

Goradia T M, Taylor R H and Auer L M 1997 Robot-assisted minimally invasive neurosurgical procedures: first experimental experience *Joint Conf. Comput. Vision, Virtual Reality, Robotics in Med. and Computer-Assisted Surgery (Grenoble, 1997)* 1st edn, ed J Troccaz *et al* (Berlin: Springer) pp 319–22

Graves B S, Tullio J, Shi M and Downs J H III 1997 An integrated remote neurosurgical system *Joint Conf. Comput. Vision, Virtual Reality, Robotics in Med. and Computer-Assisted Surgery (Grenoble, 1997)* 1st edn, ed J Troccaz *et al* (Berlin: Springer) pp 799–808

Hunter I, Jones L, Doukoglou T, Lafontaine S, Hunter P and Sagar M 1994 Ophthalmic microsurgical robot and surgical simulator *Proc. SPIE* **2351** 184–90

Kavoussi L R, Moore R G, Adams J B and Partin A W 1995 Comparison of robotic versus human laparoscopic camera control *J. Urol.* **154** 2134–6

Kwoh Y S, Hou J, Jonckheere E A and Hayati S 1988 A robot with improved absolute positioning accuracy for CT guided stereotactic brain surgery *IEEE Tran. Biomed. Eng.* **35** (2) 153–60

Lavallée S, Troccaz J, Gaborit L, Cinquin P, Benabid A L and Hoffmann D 1992 Image guided operating robot: a clinical application in stereotactic neurosurgery *Proc. IEEE Int. Conf. Robotics and Automation (Nice, 1992)* (Los Alamitos, CA: IEEE Comput. Soc. Press) pp 618–24

Maciunas R J (ed) 1993 *Interactive Image-Guided Neurosurgery* (Park Ridge IL: American Association of Neurological Surgeons)

Margossian H, Garcia-Ruiz A, Falcone T, Goldberg J M, Attaran M and Gagner M 1998 Robotically assisted laparoscopic microsurgical uterine horn anastomosis *Fertility Sterility* **70** (3) 530–34

Masutani Y, Masamune K, Suzuki M, Dohi T, Iseki H and Takakura K 1995 Computer aided surgery (CAS) system for stereotactic neurosurgery *Int. Conf. Comput. Vision, Virtual Reality, and Robotics in Med. (Nice, 1995)* 1st edn, ed N Ayache (Berlin: Springer) pp 247–51

Mettler L, Ibrahim M and Jonat W 1998 One year of experience working with the aid of a robotic assistant (the voice-controlled optic holder AESOP) in gynaecological endoscopic surgery *Human Reproduction* **13** (10) 2748–50

Mitsuishi M, Iizuka Y, Watanabe H, Hashizume H and Fujiwara K 1998 Remote operation of a micro-surgical system *Proc. IEEE Int. Conf. Robotics and Automation (Leuven, 1998)* vol **2**, pp 1013–9

Mitsuishi M, Watanabe T, Nakanishi H, Hori T, Asai R and Watanabe H 1995 A tele-micro-surgical system that shows what the user wants to see *IEEE Int. Workshop Robot and Human Commun.* pp 237–46

Partin A W, Adams J B, Moore R G and Kavoussi L R 1995 Complete robot-assisted laparoscopic urologic surgery: a preliminary report *J. Am. Coll. Surgeons* **181** 552–7

Rovetta A, Sala R, Wen X, Cosmi F, Tongo A and Milanesi S 1995 Telerobotic surgery project for laparoscopy *Robotica* **13** (4) 397–400

Schulam P G, Docimo S G, Cadeddu J A, Saleh W, Brietenbach C, Moore R G and Kavoussi L R 1997 Feasibility of laparoscopic telesurgery *Joint Conf. Comput. Vision, Virtual Reality, Robotics in Med. and Computer-Assisted Surgery (Grenoble, 1997)* 1st edn, ed J Troccaz *et al* (Berlin: Springer) pp 809–12

Schweikard A, Tombropoulos R, Adler J R 1995 Robotic radiosurgery with beams of adaptable shapes *Comput. Vision, Virtual Reality and Robotics in Medicine, First Int. Conf. CVRMed Proc.* (Berlin: Springer) pp 138–49

Slatkin, Brett A, Burdick J and Grundfest W 1995 The development of a robotic endoscope *IEEE Int. Conf. Intelligent Robots and Sys.* (Los Alamitos, CA: IEEE Comput. Soc. Press) vol **2**, pp 162–71

Taylor R H 1999 Medical robotics and computer-integrated surgery *Handbook of Industrial Robotics* 2nd edn, ed S Y Nof (New York: John Wiley)

RELATED WEB SITES

www.computermotion.com
www.ius.cs.cmu.edu/mrcas/links_geo.html (robotic surgery jump station)
robotics.me.jhu.edu/ (robots in medicine at Johns Hopkins University)
www.cs.jhu.edu/labs/cis/about.html (Computer Integrated Surgery Lab at Johns Hopkins)
www.intuitivesurgical.com/ (da Vinci Surgical System)
www.aats.org/ (search on robotic)

CHAPTER 15

ABLATION

Supan Tungjitkusolmun

Percutaneous image-guided ablative therapies using thermal energy sources, such as radio-frequency, laser, high-intensity focused ultrasound and microwave, or induction of cold (cryoablation) have received much attention as minimally invasive means for the treatment of malignant diseases in recent years. There have also been some investigations on the use of chemical ablation for various symptoms, such as cardiac arrhythmias and for removal of gallstones. In general, ablation refers to removal or elimination of a body part or the destruction of its function. This chapter discusses several types of ablation, focusing on destruction of tissue function by heating, cooling and chemical reaction. Instruments for cutting during surgery (surgical knives) and mechanical ablation are discussed in chapter 11.

Potential benefits of ablation procedures include the ability to ablate tumors in nonsurgical candidates, reduced morbidity as compared to surgery, elimination of the need for general anesthesia and shorter recovery period. In the past, a key limitation of these methods was the lesion dimensions produced for a single application of energy, for example 30 mm diameter of lesions obtained from a single conventional radio-frequency hepatic ablation probe. Because most potentially treatable hepatic tumors are larger than this, extended time for repetitive electrode placements have been required to treat all but the smallest tumors. Thus, the objective for subsequent development has been directed at enlarging the lesion dimensions produced from a single application of energy. This chapter discusses the basic concepts of the physics involved in each type of ablation, then discusses current designs and applications, limitations and future trends.

15.1 SIGNIFICANCE AND PRESENT APPLICATIONS

Gaining access to certain areas of the body is a complicated procedure which may cause excessive hemorrhage. Control of bleeding is made difficult by fragile tissue, coagulopathies, high vascular solid organs, and lesions with overlying distended blood vessels. From the patient's standpoint, surgical procedures might be criticized as associated with painful incisions, necessity for prolonged hospitalization with attendant costs, potential for infection in incisional or operative sites, or the requirement for multiple operative procedures.

Most current ablative procedures are very minimally invasive, compared to conventional surgical techniques. The procedure for a typical ablative therapy involves an introduction of a small catheter or applicator into the body, aided by an image guidance system. Once a target is located, some form of energy is delivered to the tip of the applicator which emanates thermal heat (or cooling, chemical effect) at the interface between the applicator and the tissue. Generally, the controlling parameters for lesion size are the amount of power delivered to the site and the duration of application. Table 15.1 lists some current applications of different modalities of ablation. Radio-frequency (RF) energy has been the most popular type of ablation but others are currently being explored and might eventually replace some existing applications of RF ablation due to their added advantages. For example, RF ablation is a method of choice for curing the Wolff–Parkinson–White syndrome, and atrioventricular nodal re-entry. The success rate in these cases approaches 100%. However, some cardiac arrhythmias require deep lesions that RF technologies have not been able to provide. Current research for cardiac ablation includes the use of laser, microwave and ultrasound as alternatives to RF.

Table 15.1. Examples of ablation procedures that are currently performed in clinics. C = cryoablation, US = ultrasound ablation, RF = radio-frequency, MW = microwave.

Application	Technique
Cardiology (cardiac arrhythmias)	RF
Urology (benign prostatic hyperplasia, gallbladder)	C, US, laser, RF, MW, chemical
Neurology (brain cancer)	US, RF
Oncology (tumors)	MW, RF, C, laser
Dentistry	Laser, chemical
Ophthalmology (cataracted lens, corneal sculpting, astigmatism)	Laser, US

15.2 RADIO-FREQUENCY ABLATION

In addition to being an effective cure for supraventricular arrhythmias, RF has been used to destroy tumors, such as for the treatment of liver carcinoma, hepatic and cerebral metastases, and bone lesions such as osteoid osteomas (Goldberg and Gazelle 1998). Using CT, MR or ultrasound imaging guidance, a small electrode is placed directly into the tumor. An important limitation of RF ablation has been small lesion dimensions produced by this technique. Several approaches have been explored to increase the diameter of lesion formed by the conventional RF probes. This section describes the basic principles of RF ablation, developments of the new RF technologies, and potential biophysical limitations to RF ablation.

15.2.1 Background

The principle of using alternating current during surgical procedures to avoid the undesirable effect of neuromuscular stimulation was first introduced by A J D'Arsonval in 1891. In the 1920s and 1930s, neurosurgeon H Cushing, with his engineer W T Bovie, studied the use of high-frequency current for the purpose of electrocoagulation. Subsequently, the same technique was applied for destruction of tumors. The first commercially available generators for the purpose of therapeutic controlled lesions in the brain were built by S Aranow and B Cosman in the 1950s. The technique has developed as an effective cure for various diseases.

15.2.2 Mechanisms of RF energy-induced tissue injury

The primary mechanism of tissue injury in response to RF ablation is thermal. In addition to the transformation of electric energy into thermal energy, the oscillating electromotive force may exert a direct effect on the cells. Although the thermal effect on tissue in the central zone of the RF lesion is dominant, the relative contributions of thermal and electric injury to lesion formation at the border zone is unknown (Haines 1995).

Time and temperature relationships of thermal injury to various tissues have been investigated in the hyperthermia oncology literature. Bromer *et al* (1982) reported that when human bone marrow cells in culture were exposed to temperatures of 42 °C, 55% of the cells were killed after 300 min. In contrast, exposure to 45.5 °C killed 99% of the cell population within 20 min. Figure 15.1 illustrates the importance of temperature and duration on cell killing.

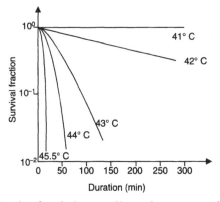

Figure 15.1. A qualitative plot of survival curves of human bone marrow cells. The survival fraction is on a logarithmic scale, while the time axis is on a linear scale. Adapted from Bromer *et al* (1982).

Figure 15.2 illustrates a heat transfer process for RF cardiac ablation and the energies involved. The electric energy is delivered from the RF generator to the metal electrode. Joule heating causes the temperature in the tissue surrounding the catheter tip to rise. Thermal conduction causes the temperature deeper in the myocardium to rise. Thermal conduction also causes the temperature in the catheter tip to rise. The blood convection in the cardiac chamber cools down the surface of the electrode and the myocardium. Other applications of RF ablation also involve a similar heat transfer mechanism. The amount of blood flow and perfusion is dependent on the anatomical location of the target.

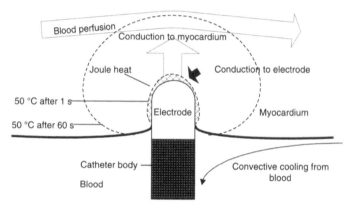

Figure 15.2. The Joule heat generated from the catheter tip elevates the temperature of the surrounding tissue. Then the thermal energy is transferred deep into the myocardium by thermal conduction and some heat is lost due to the blood perfusion and conduction to the metal electrode. Flowing blood in the cardiac chamber cools down the surface of the electrode and the myocardium.

15.2.2.1 The bio-heat equation. The bio-heat equation (15.1) governs the temperature and potential distributions in the tissue during RF ablation. Joule heating arises when energy dissipated by an electric current flowing through a conductor is converted into thermal energy

$$\rho c \frac{\partial T}{\partial t} = \nabla \cdot k \nabla T + \boldsymbol{J} \cdot \boldsymbol{E} - Q_h .\tag{15.1}$$

Since the normal body temperature is 37 °C, we have the following initial and boundary conditions:

$$T = 37\,°\text{C when } t = 0 \text{ at } \Omega, \text{ and}$$
$$T = 37\,°\text{C when } t \geq 0 \text{ at } \partial B_0$$

where Ω is the entire domain of the system, T is the temperature distribution in Ω, ρ is density, c is specific heat, k is thermal conductivity, $\boldsymbol{J}$ is current density,

E is electric field intensity, Q_h is heat loss due to blood perfusion in the ventricular wall and ∂B_0 is the outer-most boundary of the system.

In the case of cardiac ablation, we can calculate the heat fluxes at the blood–catheter, and the blood–tissue interfaces by

$$k\frac{\partial T}{\partial n} = h_b(T - T_{bl})\tag{15.2}$$

where h_b is the convective film coefficient due to the blood flow.

We can solve equation (15.1) in two steps. First, since there is no source of RF energy within the body, we can solve the Laplace equation (15.3) to find the potential distribution V

$$\nabla \cdot \sigma \nabla V = 0\tag{15.3}$$

with the following boundary conditions:

$$V = 0 \text{ at } \partial B_0, \text{ and}$$
$$V = V_0 \text{ at } \partial B_1$$

where V is the potential distribution in Ω, σ is the electric conductivity, V_0 is the applied RF voltage and ∂B_1 is the surface of the ablation electrode. Knowing the potential distribution V, we can compute the electric field intensity from

$$E = -\nabla V\tag{15.4}$$

and the current density from

$$J = \sigma E\,.\tag{15.5}$$

Then, equation (15.1) is solved for the temperature distribution T. Since the geometries of the materials in this system (tissue, ablation electrode) are complicated, researchers utilize the finite element analysis to solve the bio-heat equation and to predict the lesion size produced by different geometries of the ablation electrode, power delivered and duration (Panescu *et al* 1995, Tungjitkusolmun *et al* 2000). By solving the bio-heat equation, we can observe the extent of lesion produced by different ablation electrode designs.

15.2.3 Designs of RF ablation system

The RF field from the conducting electrode and thus the power dissipation in tissue heats the targeted tissue. The heated tissue in turn raises the temperature of the electrode tip. The ablation electrode, therefore, is designed so as not to sink away too much of the thermal energy. The lesion size is often defined as the 50 °C isotherm, which increases with rising tip temperature and increases with increasing tip dimension (Nath *et al* 1993). An optimal RF ablation electrode should be able to heat the desired tissue above 50 °C without heating the tissue above 100 °C when popping occurs.

Figure 15.3 shows a cardiac ablation electrode with a bead thermistor embedded at the tip (Edwards and Stern 1997). A thermal insulating sleeve surrounding the sensing element blocks the transfer of heat from the electrode to the thermistor. In present clinical use, the diameter of the catheter is 7 French (2.3 mm), and the length of the tip electrode is 4 mm. Figure 15.4 shows a fully deployed RF probe for use in hepatic ablation. The probe is made of a metallic needle (14–17 gauge), which houses small wire electrodes. Clinicians retract the wire electrodes into the needle during probe insertion. Once a target is reached, the four wire electrodes are deployed in order to create larger lesions. Figures 15.5(*a*) and 15.5(*b*) show the actual pictures of both devices.

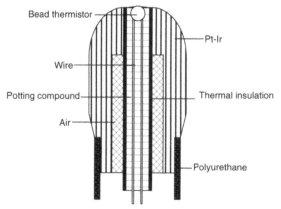

Figure 15.3. A typical ablation electrode system with thermistor embedded at the tip. A thermal insulating sleeve surrounding the sensing element blocks the transfer of heat from the electrode to the temperature-sensing element. Thus, the thermistor measures temperature without being affected by the surrounding thermal mass of the electrode. Adapted from Edwards and Stern (1997).

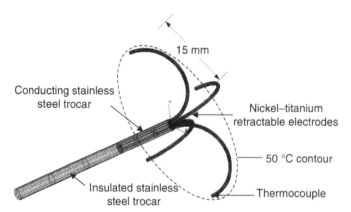

Figure 15.4. Radio-frequency probe for liver cancer. The four wire electrodes have thermocouples for temperature sensing at the tips.

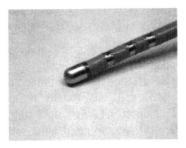

Figure 15.5. (*a*) 7F 4 mm cardiac ablation catheter (EP Technologies). (*b*) Four-tine hepatic RF ablation probe (RITA).

Figure 15.6 shows the instrumental set-up for a cardiac ablation system consisting of a catheter, a console and a ground pad that is placed on a body surface. The probe senses the temperature in the vicinity and the impedance of the system, then feeds the information back to the console. The console performs a decision test via its built-in algorithm and regulates the amount of current delivered to the probe. Generators usually produce a continuous unmodulated sine wave output at a frequency between 300 kHz and 1 MHz.

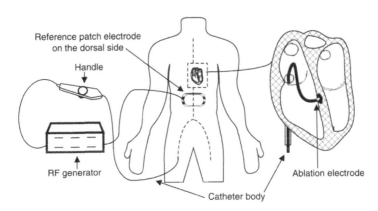

Figure 15.6. An ablation catheter is advanced into a cardiac chamber. The RF generator delivers current to the ablation electrode at the tip of the catheter. Adapted from Panescu *et al* (1995).

In addition to being able to create desired lesion dimensions, we should also consider the following operating parameters for a RF ablation system: (1) stability and consistency of the system to ensure safe operation; (2) compactness and portability; (3) durability and ruggedness. For example, a good design for a cardiac ablation system should include a steerable and flexible catheter, and the temperature-sensing element at the tip of the electrode must be well calibrated. The RF generator should be consistent, durable and include a safety mechanism to prevent overheating.

15.2.4 Advantages and limitations

The lesions generated by RF ablation are predictable, thus it can be safely applied. Furthermore, the technique is generally very minimally invasive. However, the extent of lesions produced by RF ablation has been limited due to the constraint in the size of the RF probe. Most malignant tumors are usually large, it requires repeated applications of RF ablation and longer operation time. Several ongoing researches (discussed in the next section) are dealing with this issue by modifying the technique to provide larger lesion formation.

15.2.5 Applications of radio-frequency ablation

Some of the applications of RF energy include neurosurgery, dermatological oncology, control of chronic pain syndrome, cardiology, etc. This section focuses on two of the most popular applications of RF ablation: cardiac ablation and hepatic ablation.

15.2.5.1 Radio-frequency cardiac ablation. RF ablation has become a method of choice for treatment of supraventricular tachycardias (Huang and Wilber 2000, Zipes 1994). Figure 15.6 shows a diagram of a cardiac ablation system. One example of its use is patients with refractory arrhythmias of the upper chamber who need complete interruption of the atrioventricular conduction pathway. Electrophysiologists usually monitor the abnormal conduction of the heart by advancing a few ablation catheters into the cardiac chambers, each catheter having three recording electrodes and one ablation electrode at the tip (figure 15.5(*a*)). The catheters are placed at different locations, such as the sinoatrial node, the atrioventricular node and in the ventricles. By observing the timing of the excitation activity at various locations, we can determine the locations of unwanted conduction pathways. Once the target has been localized, they can be destroyed with RF energy. The power delivered for cardiac ablation is approximately 10–50 W, and the duration is from 10 to 60 s, depending on the location of target. In high blood flow regions, such as areas above the mitral valve leaflets, we need to apply more RF power for a longer duration than in low flow areas (underneath the mitral valve).

Currently, the main challenges for cardiac ablation lie in the development of RF ablation for use in patients with atrial fibrillation and ventricular tachycardia (VT). Even though ventricular tachycardia is more serious, RF ablation has not been a highly successful technique for curing VT. In order to apply RF ablation to VT, we usually need to create a deeper and wider lesion (Panescu 1997). Conventional RF ablation electrodes often fail to produce such a lesion. The limitation of the lesion dimensions is because heating is proportional to the square of the current multiplied by the resistance (J^2R). In turn, the current density in tissue drops off as the inverse square of the distance from the electrode. As a result, heating drops off as the inverse of the fourth power of the distance. The impedance increase of the tissue adjacent to the electrode at 100 °C limits lesion size. To cure atrial fibrillation, the most prevalent form of arrhythmia, it has been suggested that long and thin lesions along trajectories similar to those created by the 'MAZE' procedures are needed (McRury *et al* 1997). Both atrial fibrillation and ventricular tachycardia represent dangerous cardiovascular conditions, and require minimally invasive techniques and devices.

15.2.5.2 Hepatic ablation. The American Cancer Society estimated that about 13,900 cases of primary liver cancer (cancer that originates within the liver) were be diagnosed in 1998 and roughly 13,000 people died of the disease. The most common form of primary liver cancer is hepatocellular carcinoma, which arises from the hepatocytes, the main type of cell within the liver. Primary liver cancer is twice as common in men than in women. Liver cancer can also occur as a secondary cancer (metastasis, meaning cancer that has spread from elsewhere in the body). Between 30 and 70% of patients who die from cancer have detectable liver metastases at autopsy.

Figures 15.4 and 15.5(*b*) show a typical RF ablation probe for liver tumor ablation. Using ultrasound for guidance, the surgeon inserts a needle probe through the skin into the liver to the area of the tumor. Once at the site, the tip of the probe is opened to expose an umbrella-like network of wires. Next, a generator device delivers RF current through the probe and wires for approximately 5–15 min. The heat generated by the RF energy kills the tumor cells. The probe measures the electrical impedance of the tissue to regulate the delivery of energy and destruction of the tumor. Over time, the body gradually absorbs the dead tissue, so surgical removal is not necessary. The current probe design is capable of destroying an area of approximately 3 cm in diameter. A smaller probe is available for much smaller tumors. For larger tumors, the surgeon has to perform multiple ablations by repositioning the probe until the entire area has been treated. However, this process is inconvenient and time consuming. The reason for the limitation of lesion size is similar to the situation in cardiac ablation. Larger needle sizes only produce slightly larger lesion diameter, and the maximum lesion diameter does not exceed 3 cm for practical probe sizes. Several investigators have used central cooling of the electrode with perfusion to increase lesion size (Goldberg *et al* 1998). Rossi *et al* (1996) reported successful treatments of hepatocellular carcinomas using bipolar electrodes for RF ablation.

15.2.6 Research

Because the goal of tumor eradication necessitates ablating the entire unwanted tissue and a small peripheral margin of normal tissue, complete ablation of the entire tumor often requires the induction of large volumes of lesion. Several recent technical innovations, including RF application to multiple and/or internally cooled electrodes enable increased energy deposition into tissues with a resultant increase in the volume of induced lesion (Goldberg *et al* 1996, Lorentzen *et al* 1997). Goldberg *et al* (1998) report that a simultaneous pulsed-RF application to a cluster of three closely spaced internally cooled electrodes enables larger volume necrosis in *ex vivo* liver, *in vivo* tissues and hepatic colorectal metastases than a conventional electrode for a single application of RF technique. Other novel techniques include bipolar arrays (McGahan *et al* 1996) and saline injections during RF cardiac ablation (Brucker *et al* 1997).

Cooling due to blood perfusion in tissue limits both tissue heat deposition and the extent of necrosis induced by RF ablation in vascular tissues and tumors. Development of strategies to reduce blood flow during RF tumor ablation may allow for improved treatment efficacy. Further study is necessary to determine under which conditions a particular method will prove superior to others, and to determine whether any of the new methods can improve patient outcomes (Goldberg and Gazelle 1998). Other topics that are currently under investigation include the effect of electrode–tissue contact and blood stream on lesion making; the effect of the electrode geometry and temperature on the lesion size.

15.3 LASER ABLATION

'*Laser*' stands for light amplification by stimulated emission of radiation. The uniqueness of lasers derives from a number of special properties, such as high intensity, coherence (photon waves are all in phase, which permits focusing on a spot size of 1 μm or less), collimation, extreme monochromaticity (photons stimulate photons of the same frequency), and the ability to be delivered in very brief pulses. The ability to launch high optical power into single optical fibers extends enormously the clinical applications of lasers, since it permits remote delivery (endoscopic, intravascular, intracavitary, interstitial) of precisely controlled therapeutic energy to tissue deep within the body with minimal damage to intervening structures (see chapter 11). Laser ablation has found wide application in ophthalmology and dentistry. This section provides a basic understanding of lasers and discusses their applications in ablative therapies.

15.3.1 Background

The theoretical foundation for the quantum optical process was developed by Max Planck in 1908 and later by Albert Einstein in 1917. The first laser, based on solid-state chromium crystal, was developed in 1960. The first gas laser (helium–neon) was developed by Ali Javan, William R Bennett, Jr, and D R Herriott in 1961. Johnson and Nassau developed the first crystal neodymium:yttrium aluminum garnet laser (Nd:YAG), emitting energy in the near infrared portion of the spectrum, in 1961. The argon laser, emitting energy in the blue-green portion of the spectrum, was developed by R Bennett, Jr, and his colleagues in 1962, and Patel *et al* (1964) developed the first CO_2 laser in 1964. Excimer (excited dimer) lasers were introduced in the 1960s. They are a family of high-pressure, pulsed gas lasers which produce intense ultraviolet (UV) light with high efficiency and high peak power at several wavelengths. The most common wavelengths are at 193, 248, 308 and 351 nm. The source of the commission is a fast electrical discharge in a high-pressure mixture of rare gas (Ar, Xe, Kr) and a halogen gas.

Since then, multiple laser sources have been developed. Currently, many materials (solid, liquid and gas) are used as the active medium in lasers. Laser wavelengths cover the entire visible portion of the electromagnetic spectrum as well as wavelengths in the UV and IR portions (figure 15.7). However, the only lasers for practical clinical use are the argon (Ar), CO_2 and Nd:YAG. Presently, certain lasers are considered the tool of choice for various ablative procedures. For example, laser ablation using the argon ion laser has found wide application in ophthalmology, where the optical transparency of the eye in the visible wavelength range permits precise focusing of the beam onto the target of interest. By this technique, conditions such as angioproliferative retinopathy, senile macular degeneration and retinal detachment can be safely and effectively treated. CO_2 and Nd:YAG lasers are also of value in general surgery (see chapter 11), where the ability to achieve coagulation while cutting can save time and limit blood loss. Lasers presently have more limited, but increasing roles in orthopedics, neurosurgery, and other medical and surgical specialties (Parrish and Wilson 1991).

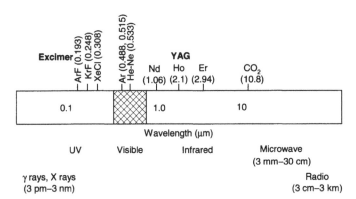

Figure 15.7. The electromagnetic spectrum is shown. The wavelengths of lasers range from ultraviolet to infrared.

15.3.1.1 Fundamentals of lasers. Quantum theory predicts that atomic or molecular systems can exist in different quantum energy states. In an atomic system, an electron can be raised from the *ground state* to an *excited state* by *absorption* of electromagnetic radiation (figure 15.8). An electron in the excited state eventually returns to the ground state, resulting in *spontaneous emission* of electromagnetic radiation of a particular wavelength. There are certain elements where instead of spontaneous emission, *stimulated emission* may occur. A photon with a wavelength identical to the absorption of the atomic system can stimulate the return of an excited electron to the ground state. In so doing the photon creates a new identical photon. By repetition of this process the two photons can now result in four identical photons. This is the process of amplification. For stimulated emission, the active medium of the laser (liquid, solid or gas) must undergo a population inversion, i.e. the number of electrons in the excited state must exceed the number of electrons in the ground state. All lasers require a pumping system to achieve the population inversion.

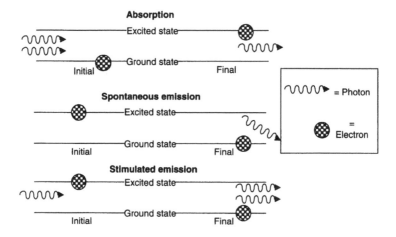

Figure 15.8. An electron can travel between the ground and excited state. The three mechanisms are absorption (electron absorbs energy and rises from the ground state to the excited state), spontaneous emission (electron in the excited state returns to the ground state and emits some energy), and stimulated emission (a photon stimulates the return of electron to the ground state).

15.3.1.2 Pulsed ablation. Although there are many applications of laser ablation using continuous wave (CW) sources, there is a growing interest in the use of short duration (less than a few ms) pulses. Laser sources generating short light pulses can produce a considerably greater range of laser–tissue interactions than CW or long pulse sources, since they can initiate nonlinear processes that alter the interaction mechanisms at work (Parrish and Wilson 1991). The efficiency of tissue ablation or photochemical reactions (see section 15.3.2) can be greater with pulsed sources than with CW sources. The advantages of using pulsed lasers are delivery of sufficient energy to ablate tissue with each pulse, which removes the hot tissue before heat is transferred to surrounding tissue (Welch *et al* 1991), and controllability of the tissue effect. For microscopic spatial confinement of thermal and mechanical effects, pulsed lasers are required (Parrish and Wilson 1991).

15.3.2 Laser–tissue interactions

When light is absorbed, the absorbing medium is increased in energy content by the energy contained by the total number of photons that were absorbed. This usually results in increasing the temperature of the absorbing medium. The relationship between the absorption and transmission of light by any homogeneous, isotropic medium obeys the Beer–Lambert Law:

$$I = I_0 e^{-\alpha x} \tag{15.6}$$

where I is the beam intensity at any given depth and I_0 is the initial beam intensity. The absorption coefficient α represents the decrease in beam intensity due to absorption. Figure 15.9 shows that the beam intensity decreases exponentially with tissue depth x. The mean free path of the beam in tissue is defined as that depth where the beam intensity has dropped to 37% (1/e) of the initial beam intensity I_0. This occurs at depth $X_e = 1/\alpha$. The CO_2 laser has a high α, thus it is ideal for vaporization of tissue with very little surrounding tissue coagulation. Thus, it has been employed for use in a scalpel.

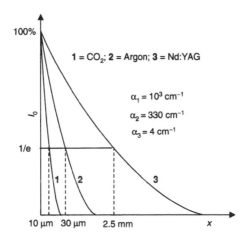

Figure 15.9. The beam intensity decreases with tissue depth. The mean free paths of CO_2, argon and Nd:YAG are 10 μm, 30 μm and 2.5 mm, respectively. α_1, α_2 and α_3 are the absorption coefficients of CO_2, argon and Nd:YAG in blood. Note that this figure is not drawn to scale.

For the Nd:YAG laser, the mean free depth (2.5 mm) is much larger than that of CO_2 (10 μm), thus it is suitable for tissue coagulation necrosis since it can generate larger lesions. For the Ar laser, its mean free path in blood is approximately 30 μm. This makes it less ideal for cutting than the CO_2 laser, and less ideal as an ablator than the Nd:YAG laser (Svenson *et al* 1987).

Although the mechanisms of laser–tissue interaction are still being studied, the following sections describe some of the known mechanisms that can happen during laser–tissue interaction.

15.3.2.1 Photocoagulation. The most widespread surgical applications of lasers in medicine employ the phenomenon of *photocoagulation*. In this process, proteins, enzymes and other critical biological molecules in tissue are heated to temperatures well above 50 °C with a resultant denaturation occurring almost immediately. This type of thermal injury of tissue is highly dependent upon the rate of energy delivery. If energy is delivered to tissue at a very high rate, a high temperature may result, which produces a more rapid rise in temperature and a more rapid lesion formation, until a tissue temperature of 100 °C is exceeded. At a temperature exceeding 100 °C, photovaporization results. Thus, to obtain optimum photocoagulation, the laser beam delivery rate must achieve a temperature in the coagulation zone of 50–100 °C. At relatively low temperatures near 45–50 °C, it takes several seconds to achieve photocoagulation, whereas at higher temperatures approaching 100 °C, photocoagulation can occur within fractions of a second. Similar to equation (15.1), the bio-heat equation for laser ablation is

$$\rho c \frac{\partial T}{\partial t} = \nabla \cdot k \nabla T + S - Q_h \tag{15.7}$$

where S (W m^{-3}) is the heat of laser ablation. The boundary condition in equation (15.2) also applies in laser ablation.

Photocoagulation is used to prevent blood loss when surgically incising heavily vascularized tissue (e.g. the liver, the vaginal wall, the larynx, etc). Photocoagulation is also used to stop gastric bleeding and, on an investigational basis, to connect severed vessels and weld tissues. In ophthalmology, the Ar laser is used to coagulate the retina, the trabecular network and the ciliary body of the eye.

Longer exposure durations are utilized to coagulate (or ablate) larger tissue volumes by making use of heat conduction during the exposure. When this technique is employed, the exposure rate must be reduced so as not to exceed the critical temperature of 100 °C. Larger coagulation zones can also be achieved through the appropriate choice of wavelength. Penetration of energy can be achieved directly by selecting a wavelength with a deeper penetration depth. For example, Nd:YAG has a mean free path of 2.5 mm, while CO_2 has a mean free path of 10.6 μm in tissue. However, when large lesion formation is occurring, the tissue surface may not appear to change as much as when a nonpenetrating wavelength is chosen. Hence, the surgeon must be careful when employing the Nd:YAG laser to ensure that the coagulation zone is not too large.

15.3.2.2 Photovaporization. Tissue vaporization occurs at temperatures in excess of about 100 °C when the tissue water boils. The production and emission of steam results in the removal of biological tissue in the form of microscopic particles. Photovaporization is used for incision and removal of diseased tissue. In this process, the exposure rate must be sufficiently high to produce a surface removal without photocoagulation of too large a tissue volume of adjacent tissue. The boiling off of surface tissue and the rate of delivery of optical energy determines the peak temperature rise that, if properly controlled, minimizes the area of undesired coagulated adjacent tissue. The system for photovaporization usually includes fume extractors to evacuate the particles produced in this process unless the volume of tissue is extremely small. Chapter 11 describes applications of photovaporization of lasers.

15.3.2.3 Photochemical ablation. Photochemical ablation is a procedure that is associated with low laser exposure rate that does not produce a significant temperature increase in the irradiated tissue, but does interact with a natural photosensitizer to produce the desired reaction. An example of photochemical ablation in medicine is the treatment of cancer by injection of *hematoporphyrin* derivative and irradiation with a red light. This process is also useful for clean-cut incisions that are possible with short-wavelength UV pulsed excimer lasers. As the incident laser radiation is shifted toward longer wavelengths (lower photon energies), the photochemical ablative component diminishes and the photothermal mechanism dominates. One of the applications of photochemical ablation is corneal shaping using a pulsed 193 nm excimer laser, since the 193 nm photons are absorbed by protein in the tissue as well as by cellular DNA. The depth of penetration of the incident photons determines the volume removed with each pulse of laser energy (Welch *et al* 1991).

15.3.3 Advantages and limitations

Laser ablation can potentially create large lesions, although this ability may be limited by the decrease in thermal conductivity that occurs with desiccation of the catheter–tissue interface. Some of the potential advantages in laser ablation are:

1. ability to coagulate, or vaporize by varying the power and exposure time of application;
2. multiple wavelengths with selective absorption by various tissues;
3. selective tissue ablation by modification of pulse characteristics and photonsensitizers;
4. less operative and postoperative pain.

Some of the disadvantages in application of the laser for ablative therapy are:

1. high cost of equipment;
2. lack of portability of instrumentation;
3. certain technological limitations, such as the specific, perpendicular requirement for fiber orientation and the unpredictable photon scattering caused by nonuniform tissue optical properties, may create difficulties in controlling lesion size and geometry;
4. hazards to patient and personnel from inadvertent laser exposure, electrical injury or fires.

Safety is a vital factor common to all laser procedures. The patient and the operator must be protected properly. Study and utilization of current national guidelines, such as the American National Standard Institute Z136.1 and Z136.3 in conjunction with current medical publications, provides a foundation for laser safety for any laser practitioner (Sliney and Trokel 1993).

15.3.4 Applications

Laser ablation of the eye has been the most common application. It is currently also used to destroy malignant tumors, lung cancer, in arthroscopic procedures, vocal cord paralysis, arterial blockages, dental material, and angioplasty of perpheral and coronary arteries. Applications of lasers for endoscopic surgery are presented in chapter 11. We discuss some of the most common laser ablative therapies in this section.

15.3.4.1 Ophthalmology. Lasers have been used for removing cataractous lens tissue, curing astigmatism and corneal correction (see chapter 11). In figure 15.10, a patient's eye is shown with a cataracted natural lens. The vitreous membrane separates the vitreous body from aqueous contents in the anterior chamber. Field viewing via objective lens is through a partially reflecting mirror, which also serves to focus the bundle of rays shown issuing from a fixed objective lens D and convergent at a desired angle to the surface of the cornea (*x*- and *y*-axes manipulation). The laser beam on the right is expanded via the first pair of lenses A and B, and the second pair (lens C and D). The beam is collimated between lens B and C. Lens C is mounted for axial displacement of *z*-axis manipulation of the depth of the focal spot of the laser on the cataracted natural lens. The *laser spot projection* shown in figure 15.10 is a steady source of visible light (e.g. He–Ne). The light source also travels through the focusing system of lens C and D, and it is a visible spot in the field of view of the viewing microscope or camera. Pulsed Nd:YAG lasers and UV excimer lasers can be used as the source in figure 15.10. The energy of individual pulses is about 1–30 mJ.

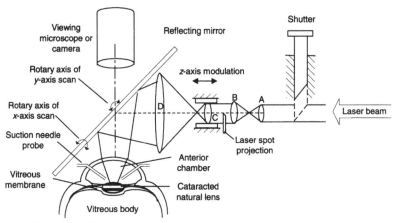

Figure 15.10. A simplified optical diagram of components of the laser system for removing cataracted lens tissue. From L'Esperance (1985).

The procedures involved during surgery are:

1. The surgeon monitors the patient's eye through the viewing microscope (or camera).
2. He then finds a focal spot using the laser-spot light projection, and positions the focal spot at a point on the outer perimeter of the desired limit of surgery (by controlling the joystick and the z-axis driver knob in figure 15.11 manually). The location of the focal spot is stored in the limiting coordinate storage.
3. He then moves the joystick along a circular path for the limiting perimeter of lens–tissue decomposition, and the data are once again stored in the limiting coordinate storage.
4. The surgeon then centers the focal spot, and can start the ablation procedure.

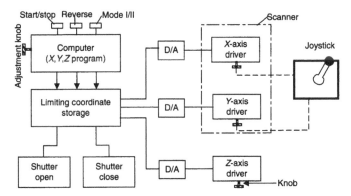

Figure 15.11. A simplified block diagram of computer and scanner control elements used in connection with the laser instrument shown in figure 15.10. D/A is a digital/analog device (can be digital-to-analog converter and analog-to-digital converter). From L'Esperance (1985).

15.3.4.2 Tumor surgery. Lasers have also been used in oncologic surgery to remove tumors by either excision or direct ablation. Lasers have been demonstrated to have an advantage in removing tumors that are in difficult-to-access areas, and can remove tumors in heavily vascularized tissue by minimizing blood loss during such surgical procedures. The laser has also shown its value for ablating or surgically incising tumor tissue in the brain where microsurgical techniques are highly desirable. In these applications, the most conventional laser employed has been the CO_2 laser.

15.3.4.3 Laser for catheter ablation. Because of the brief, intense exposure and high absorption coefficients of Ar, CO_2 and excimer lasers there is rapid heating of a small tissue volume and very little heat transfer to surrounding structures. Thus, these lasers are not well suited for percutaneous arrhythmia ablation. The Nd:YAG laser operates at a wavelength of 1064 nm. This is within the optical window and therefore is associated with a relatively low absorption coefficient, and high tissue penetration depth. When used at relatively low exposure rate for a long duration, these characteristics allow the Nd:YAG laser to produce heating and photocoagulation of a region of myocardium significantly larger than the point of contact. These biophysical properties make the Nd:YAG laser well suited for ablation of arrhythmogenic sites within the heart. Nd:YAG irradiation has been used for resection of intraoperative ablation of ventricular tachycardia in patients with coronary disease and history of myocardial infarction. The irradiated myocardium was free of overlying blood during cardiopulmonary bypass. A power output of 20–80 W was applied through an optical fiber of 0.6 mm in diameter. Large regions of myocardium could be ablated from both the epicardial and endocardial surface using this technique. However, there was an overall high perioperative mortality rate. Ar laser ablation has also been used as an adjunct to operative resection of ventricular tachycardia.

Weber *et al* (1989) presented a clinical data experiment with percutaneous laser catheter ablation (continuous wave Nd:YAG). They delivered radiation at a power of 10 W for 3, 5 and 10 s through a 400 μm optical fiber. The diameter of the lesions was approximately 7 mm and the depth was 11 mm. The complexity of the energy source and of the delivery system represents a disadvantage of this technology. The ability to effectively ablate arrhythmogenic myocardium is limited by the need to exclude blood from the catheter tissue interface. The potential to volatilize the target also raises safety issues with regard to myocardial perforation.

15.3.4 4 Urology. Removal of lesions and tumors of the bladder and urethra are all current applications of lasers by urologists. Ar, CO_2 and Nd:YAG lasers have all been used in urologic surgery with both fiber-optic and cystoscopic delivery systems.

15.3.4.5 Neurosurgery. CO_2, argon and Nd:YAG lasers have been used in microsurgery to selectively ablate tissue or to remove tumor tissue that is extremely close to critical neural tissue that must not be disturbed. The use of pulsed modes of operation to limit the heat conduction is generally necessary for microsurgery (Jeffreys 1992).

15.3.5 Current research

Some of the ongoing researches on laser ablation are:

1. *Laser angioplasty.* The use of lasers to ablate plaque in coronary vessels has been a major challenge in laser medicine. There are many disparate approaches to this complex problem, involving different lasers (argon-ion, excimer, excimer-pumped dye, holmium: CW or pulsed), different delivery systems (single or multiple optical fibers, laser-heated metal tips, balloon catheters with laser heating) and different control and guidance methods (optical or ultrasonic imaging, fluorescence spectroscopy and imaging) (Parrish and Wilson 1991).
2. *Reshaping the cornea.* This is done using 193 nm ArF excimer laser ablation in order to correct refractive abnormalities of the eye. This represents a medical procedure that depends critically on the ability of the laser to deliver controlled energy with good precision.
3. *Ultrashort pulsed laser.* Capitalizing on the evolving technology of ultrashort pulse lasers could result in many advantages for biomedical applications. The major advantages of the ultrashort pulse laser (USPL) tissue ablation method are:

    a. efficient ablation due to the small input of laser energy per ablated volume of tissue and the resulting decrease of energy density needed to ablate material;
    b. the ablation threshold and rate are less dependent on tissue type and condition;
    c. high precision in ablation depth is achievable because only a small amount of tissue is ablated per pulse;
    d. low acoustical noise level (as compared to acoustical noise produced by other laser systems);
    e. minimized pain due to localization of energy deposition and damage.

15.4 ULTRASOUND ABLATION

The ability of ultrasound to produce small foci of high energy density opens up the possibility of high-temperature, short-duration hyperthermia and tissue ablation. These techniques for thermal therapies are attracting considerable research, clinical and commercial interest. Ultrasound has relatively large values for the ratio source dimensions/wavelength for transducers of practical size, enabling collimated and focused beams to be produced. It also has a relatively low attenuation in soft tissues (up to 1 dB cm^{-1} MHz^{-1}) for frequencies of interest (between 0.5 and 10 MHz). We discuss some of the basic fundamentals of ultrasound and its applications in this section.

15.4.1 High-intensity focused ultrasound: background

High-intensity focused ultrasound (HIFU), in the frequency range of 500 kHz to 10 MHz, has been used in a wide range of therapeutic applications. Fry *et al* (1954) first developed this technique as a method of selective brain tissue destruction for neurosurgical research. The following years have seen a steady progress in understanding the basic physical principles involved in the interaction of this mechanical vibratory waveform and biological tissue. Developments in transducer technology have been helpful in recent device designs, which are potentially useful in the treatment of focal disease in humans (Fry 1993).

15.4.1.1 The ultrasound source. The initial work with ultrasound sources used materials such as quartz, lithium sulfate, or barium titanate. Today new powerful piezoelectric materials are available. Depending on the application, one or other material is more advantageous due to physical or cost factors, or simply because of a less complicated manufacturing process. Figure 15.12 shows a simplified arrangement of sound source, focal beam configuration, transmission path and focal site dimensions.

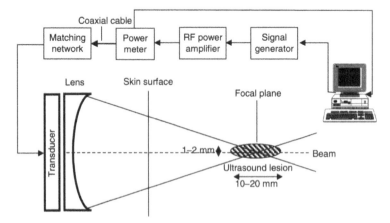

Figure 15.12. A radio-frequency signal is produced by a signal generator and amplified by a RF power amplifier. A power meter is used to monitor the forward and reflected power in the coaxial cable connected to the transducer's matching network.

15.4.1.2 Intensity, attenuation, absorption and scattering. Some of the parameters used to characterize ultrasound are acoustic intensity, attenuation, absorption and scattering. The acoustic intensity I (W m^{-2}) is the rate of energy flow through a unit area normal to the direction of wave propagation. For a plane continuous wave, the temporal average intensity is

$$I = \frac{p_0^2}{2Z} \tag{15.8}$$

where p_0 is the amplitude of the acoustic pressure and Z ($= \rho_0 c$) is the acoustic impedance. c is the speed of sound in the tissue and ρ_0 is the equilibrium density. Ultrasound energy attenuates when propagating through the tissue. The relationship between I_0, the intensity at the surface, and I_z, the intensity at a depth z for a plane wave, is

$$I_z = I_0 e^{-2\alpha z} \tag{15.9}$$

where α is the amplitude attenuation coefficient. The attenuation coefficient consists of two parts, one (α_a) due to absorption and the other (α_s) due to scattering:

$$\alpha = \alpha_a + \alpha_s . \tag{15.10}$$

Energy absorption of tissue leads to local heating. The scattered energy also contributes to heating caused by the ultrasound field (Duck *et al* 1998). The bio-heat equation (15.7) also applies for ultrasound ablation whereas in this case ultrasound acoustic power is substituted for S.

Figure 15.12 shows a schematic of ultrasound ablation technique. A source of ultrasound is placed at a distance from the target tissue volume, and may be outside of the body or intracavitary (for example in the rectum in the case of benign prostatic hyperplasia). The focus of the beam is set to be within the tissue volume to be treated. The source is coupled to the patient by an enclosed volume of degassed water. By using high acoustic powers for short exposure times (1–2 s), ablative temperatures are achieved rapidly, and the effects of heat dissipation by blood perfusion and thermal conduction become minimal. Current transducers for HIFU are made of piezoelectric ceramic, which can be shaped and driven as phased arrays (see section 8.2.3).

15.4.2 Advantages and limitations

An important advantage of ultrasound ablation is its ability to heat larger volumes of tissue due to penetration of ultrasound energy in tissue and control of heating along the length of the applicator by controlling the power to individual transducers in multi-element configurations (phase arrayed). We can also utilize ultrasound to provide direct imaging of the ablation site. In addition, ultrasound has a relatively low attenuation in soft tissues. One of the current limitations of ultrasound has been inadequate power delivery from a small probe required in certain applications such as cardiac arrhythmias.

15.4.3 Applications

The two commercially available HIFU devices are trans-rectal devices designed primarily for urological treatments, and extracorporeal systems designed primarily for treatment of pathology in the abdomen and breast (Duck *et al* 1998). Applications of ultrasound ablation have also been successful in treatment of glaucoma (Silverman *et al* 1991).

15.4.3.1 Urology. Benign prostatic hyperplasia (BPH) is a benign condition and is usually referred to as *prostate enlargement.* BPH is not cancer and the majority of men with symptoms of BPH do not have prostate cancer. BPH is widespread, yet generally not life-threatening and most commonly affects men over 50 years of age. Approximately 11 million men in the United States currently suffer from moderate to severe BPH that requires treatment. The most popular current treatment, transurethral resection (TURP), requires general anesthesia, and is associated with significant morbidity and mortality rates.

Transrectal ultrasound is usually the imaging modality of choice for urologists who wish to investgate the prostate. It is therefore an appropriate step to therapy in such an examination. In the urological field, HIFU has been most widely used for the treatment of BPH. Foster *et al* (1993) used an applicator with a 4 MHz crystal for both imaging and ablation, and it had a focal length of 24 mm. The lesion was 7 mm long and 0.6 mm in diameter for an intensity of 16.8 MW m^{-2}. Gelet *et al* (1993) used a retractable 7.5 MHz imaging transducer, and a 2.25 MHz transducer for ablation, with a focal length of 35 mm, intensities between 7 and 10 MW m^{-2}. The lesion diameters found were approximately 1.5 mm, and the lengths of the lesions were 8 mm.

15.4.3.2 Neurology. Lynn *et al* (1942) attempted to put lesions in the brain through the skull. Acoustic absorption and mode conversion in the bone resulted in profound damage to the scalp with only small lesions being seen in the brain, mostly at entry site. Fry and Fry (1960) treated patients with Parkinson's diseased with HIFU following craniotomy (surgical removal of a portion of the skull). A general anesthesia was required. The treatment took up to 14 h. Fry and Fry (1960) reported that the symptoms associated with the Parkinson syndrome were eliminated. This technique is not used widely because it is invasive (craniotomy is required).

15.4.4 Research

Current research on ultrasound ablation involves improving the efficacy of the technique for use in the applications discussed in section 15.4.3. In addition, ultrasound has been studied for cardiac ablation. The difficulty of utilizing HIFU for cardiac ablation is the inadequate power delivery from small ultrasound transducers. To overcome this limitation, Zimmer *et al* (1995) introduced a transducer mounting configuration that improves the power output capabilities of the piezoelectric plate. This translates to relatively high conversion efficiency and large power output. They suggested that an adequate source intensity for cardiac ablation should be around 10–15 MW m^{-2}. The diameter of the catheter was 2.3 mm (7 French). When using ultrasound intensities around 150 and 300 kW m^{-2}, the *in vivo* lesion depths were between 5 and 9 mm. The deepest lesions for any given intensity were predicted to occur at approximately 10 MHz. Further experimental tests are needed to verify the potential advantages of ultrasound ablation.

15.5 CRYOABLATION

Cryoablation is a technique that generates subzero temperatures within an ice ball which has been formed by a circulating agent (cryogen) such as liquid nitrogen (LN$_2$) or argon. The technique has evolved considerably since first attempted in the 1960s. Accurate monitoring and control of the procedure is currently achieved by employing both intraoperative ultrasound and temperature monitoring. We discuss the development, basic principles and applications of cryoablation in this section.

15.5.1 Background

Treating diseases or injury through the application of cold was first recorded in 1824. In the late 1960s, investigators first began experimenting with cryoablation for localized prostate cancer. In this era, cryoablation was performed via an open incision because ultrasound technology had not evolved to aid in placement of the cryoprobes. Only one probe could be placed at a time and the proper probe position was determined by digital rectal examination. Thus, the procedure was time consuming and not well controlled. In addition, the early technology did not allow the LN$_2$ to remain in the liquid phase throughout the length of the probe. The conversion of liquid to gaseous nitrogen limited the ability to kill prostate tissue optimally at the negative temperatures required for lethal injury. Due to these limitations, application of cryoablation for the prostate was virtually abandoned in the late 1970s.

The reintroduction of cryoablation for prostate cancer occurred in the late 1980s (Onik *et al* 1988). This renewed interest was stimulated by several important technological developments. The development of a new cryogenic system allowed continuous cooling of LN_2 at -180 to -195 °C from the LN_2 to the tip of the cryoprobe. Argon, helium, hydrogen and oxygen have also been used in cryogenic devices. It is also possible to use five cryoprobes simultaneously, which reduced the procedure time. Advancements in ultrasonic imaging (see chapter 8) have given the urologists greater control of the technique.

15.5.2 Mechanism of tissue damage

Cryoablation involves freezing the water in tissue. The process of freezing is a surface phenomenon in which water molecules from the liquid water phase attach to the crystalline structure of the water molecules in ice. Since the energy of the water molecules in a liquid state is higher than the energy of the water molecules in a solid phase, energy must be removed from the water molecules that attach to the ice crystal during freezing.

Cell death happens once the tissue temperature is lower than -20 °C. Tissue damage caused by freezing occurs at several levels: molecular, cellular and whole tissue structure. Both chemical and mechanical processes attribute to the destruction of tissue during cryoablation.

Factors that affect the size of lesions during cryoablation are:

1. cold temperature produced in the tissue;
2. duration of freezing;
3. freezing rate;
4. thawing rate.

The rate of flow of a cryoprobe determines how efficiently it freezes. The greater the flow of liquid nitrogen through a cryoprobe, the greater the heat-extracting capability it has. Typical flow rates available on cryoablation systems are 25%, 50%, 75% and 100%. The following section describes some of the characteristics of the iceballs when these factors are varied.

15.5.3 Designs of cryoablation systems

Cryoablation systems can be divided into two categories: gas systems and liquid/gas systems. Each system has two components: the cryoprobe, with its associated supply lines, and the console, which supplies the cryogen. Figure 15.13(*a*) shows different sizes of argon-based probes (Cryoprobe, Endocare Inc., Irvine, CA). Figure 15.13(*b*) illustrates the Cryocare™ system (Endocare). More information on these systems is available at http://www.ecare.org.

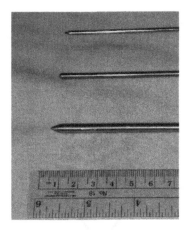
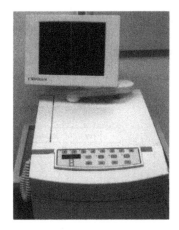

Figure 15.13. (*a*) Endocare's cryoprobes. (*b*) The argon-based eight-probe Cryocare™ system. Physicians can set the flowing rate, ablation duration, and the thawing rate, and apply up to eight cryoprobes simultaneously.

15.5.3.1 The cryoprobes. In the past, gas cryogenic systems, such as freon, nitrous oxide and CO_2, had limited freezing capacity at approximately –60 to – 80 °C (Verdier 1992). Presently, the Cryocare™ system of Endocare Inc. is a gas-based argon system that has been widely adopted in clinical settings and they produce a low temperature of –186 °C. Gas cryoprobes require no vacuum insulation as in the liquid/gas systems. This simplifies the instrument and allows much smaller probe handles, enabling multiple probes to be placed closer together.

Liquid nitrogen (LN_2) has been a widely used system in treating malignant diseases. Liquid nitrogen at its phase change temperature of –196 °C is an excellent coolant, relatively inexpensive, and easy to obtain and handle. Other cryoprobes such as liquid helium are being investigated. Liquid helium, however, is expensive and very difficult to handle because of its volatility. The leading systems based on LN_2 are the CMS AccuProbe® systems, Cryomedical Sciences, Inc. (http://www.cryomedical.com, Rockville, MD).

The LN_2 probe consists of a closed-end tube with two tubes concentrically arranged within it. Figure 15.14 shows a basic design for a typical LN_2-based cryoprobe. One tube welded to the inner aspect of the cryoprobe creates a space between it and the outer tube that can be evacuated, creating a vacuum insulation space. A variable portion of the closed-end probe is left without vacuum insulation and is the active freezing zone of the probe. The innermost tube is the supply tube for the LN_2. When the nitrogen exits the supply tube, it hits the warm uninsulated tip of the cryoprobe, where it changes phase, expanding 700 times in volume. The expanding gas exits the cryoprobe around the supply tube. This gas expansion is the constraint on the probe's functioning, since it creates a back pressure that limits the flow of liquid nitrogen into the cryoprobe. This phenomenon has been one of the most important factors to overcome in developing a smaller cryoprobe that could still freeze large volumes of tissue (Verdier 1992).

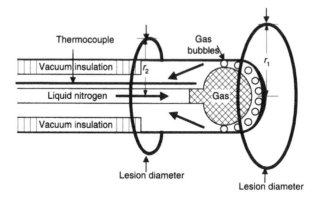

Figure 15.14. Internal structure of a typical cryoprobe. LN_2 cryoprobes must have vacuum insulation to prevent freezing up the shaft of the cryoprobe and subsequent destruction of normal tissue. The LN_2 changes phase when it hits the warm metal surface of the probe tip. Thus, a thin film of gas bubbles is formed on the metal surface.

Another phenomenon, caused by the change of phase from liquid to gas, is called *Liedenfrost boiling*. When the LN_2 hits the warm metallic tip, gas bubbles form between the liquid and the metal and they act as an insulator. Therefore, when the LN_2 changes phase at $-196\ °C$, the temperature of the cryoprobe tip is $-160\ °C$ because of this effect. Cryomedical Sciences, Inc., Rockville, MD, developed a system that utilizes LN_2 slush at approximately $-206\ °C$ to supercool the LN_2 passing through the cryoprobe. The flow of LN_2 at $-206\ °C$ would flow through the cryoprobe and exit it with a temperature below $-196\ °C$. In this situation the marked volume expansion of LN_2 into a gas would be greatly reduced and would allow substantially greater flow of cryogen into a smaller probe (as small as 3 mm in diameter). The heat-extraction capability of LN_2 is approximately 88 times more efficient than nitrogen gas at the same temperature.

Another important component of the cryoprobe is the thermocouple that determines whether the cryoprobe is working properly. If the thermocouple, which is placed near the tip, shows a temperature reading of above $-160\ °C$, the cryoprobe is not functioning properly. The thermocouple also provides the information for a temperature feedback loop, which controls the function of the cryoprobe. In certain circumstances, such as the beginning of the surgical procedure, the surgeon only wants to stick the cryoprobe to the tissue and he or she places the remaining cryoprobes (set temperature $\sim -70\ °C$). Thus the surgeon does not want the cryoprobe to function at full capacity.

15.5.3.2 The console. Basic features for the cryosurgical console include:

1. outlet for multiple probes (at least five for prostate surgery);
2. a proper discharge system for the exhaust line;
3. sufficient source for liquid nitrogen that can supply multiple probes;
4. display and menu for controlling the flowing rates, thawing rates, ablation durations, etc.

15.5.3.3 Characteristics of the cryosurgical iceball. When a higher flow rate is set, the radius r_1 at the tip and the radius r_2 near the vacuum insulator are bigger than that of slower flow rates (see figure 15.14). A typical iceball formed during cryoablation has a cone-like shape, with a larger radius at the probe tip.

A typical cryoablation system allows simultaneous transport of LN_2 to five cryoprobes. The temperature at the tip of each probe can be regulated by varying the rate of flow of LN_2. The radius of the iceball is determined by the rate of flow of the LN_2. The iceball is cone shaped, with the maximum radius being at the tip of the probe. The length of the iceball along the probe is approximately the length of the exposed section of the cryoprobe. Zippe (1996) reported that the edge of the iceball as seen on ultrasound is at 0 °C, and the tissue necrosis and cell death occur at –20 °C. At 100% maximum flow, this gradient is approximately 2 mm from the edge of the 20 mm radius iceball, near the tip of the probe (3.5 mm in diameter). At the proximal end of the probe (4 cm from the tip), the gradient from the rim and the –20 °C zone is 4 mm, with a radius of 13 mm. The iceball of a slower freeze at 25 % of maximum flow has a much smaller radius, and a much larger gradient between the edge and the –20 °C zone.

15.5.4 Advantages and limitations

Advantages of cryoablation are:

1. destruction of nonresectable tissues—cryoablation is able to destroy irregular tissue surfaces that are not resectable, such as in prostate cancer;
2. tissue sparing—the cryoprobe can be placed within the lesion and the lesion destroyed with minimal loss of surrounding tissue;
3. vessel sparing—large vessels retain the integrity of their elastic walls when frozen, and continue to function as blood conduits without showing evidence of blood clot, aneurysm formation or rupture;
4. nerve regeneration—since cryoablation does not completely denature proteins as does heating, peripheral nerves undergo axonal degeneration when frozen, but the nerve sheaths are left intact, permitting nerve regeneration of the axon through an unobstructed path. This phenomenon can be used to advantage in curing tumors of the pelvis or extremities, and prostate cancer;

5. relatively large freezing area for a given size of cryoprobe—for newer cryosurgical instruments, a 35 mm diameter iceball can be created with a single 3 mm probe;
6. predictable geometry—the shapes of the lesion can be spherical or cylindrical, depending on the active freezing length of the probe.

Some of the disadvantages of cryoablation include:

1. inaccurate monitoring of the freezing as it occurs;
2. delivery systems are bulky;
3. large size and inflexible probe, compared to RF ablation systems.

Since very small amounts of liquid cryogens can produce large amounts of gas, the users should follow manufacturers' safety guidelines when handling these materials. Cryogenic gases are capable of displacing air necessary for respiration and causing asphyxiation. Clinicians should never allow any unprotected part of the body to touch uninsulated pipes or vessels that contain cryogenic fluids.

15.5.5 Applications of cryoablation

Cryoablation has wide applications in urology and oncology. We describe an application of cryoblation for curing prostate cancer in this section.

15.5.5.1 Prostate cancer. The prostate is a small gland about the size and shape of a walnut. It is located just below the bladder and surrounds the urethra. The prostate is composed of glandular and muscular tissue. With increasing age these tissues continue to grow, resulting in enlargement of the prostate. Prostate cancer is the most commonly diagnosed cancer among US men and the second deadliest, behind lung cancer. In 1999, 184,500 US men will be diagnosed with the disease and more than 39,200 will die from it, the American Cancer Society estimates.

It is important to know the extent of tumor in the gland and the location of the tumor. Multiple spatially located prostate biopsies generally give the cryosurgeon an idea of which quadrants of the prostate must be covered well in the freezing process. The patient requires a general anesthetic and is placed on a warming blanket with air pillows to help control systemic hypothermia.

The next procedure is to insert a trocar needle (e.g. 18-gage 22 mm) into the bladder followed by inserting a small guide wire through the needle and coiled in the bladder. A dilator, along with a drainage catheter, is then inserted into the bladder. The bladder becomes distended. This enhances the ultrasonic contrast at the bladder. A urethral warming device is then passed over the wire into the bladder and the wire is removed (see figure 15.15).

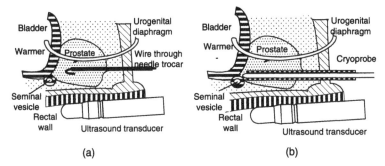

Figure 15.15 System for prostate cryoablation. (*a*) The needle trocar and the guide wire are first inserted into the body. (*b*) The cryoprobe. Adapted from Zippe (1996).

Once the proper sites for placements of the probes have been determined, the cryosurgeon places the probes so that the cylinders of frozen tissue overlap and freeze the entire prostate volume without injuring the surrounding structures, such as the rectum, bladder trigone, intramural ureter and the urogenital diaphragm. Next, a J-wire is inserted through the needle into the prostate (see figure 15.15(*a*)). The needles are then withdrawn from the prostate, leaving the J-wires in place. Then, the sheaths are advanced into the prostate at the proper locations. Finally, the cryoprobes are inserted. The probes are then activated by circulating the liquid nitrogen at a slow flow rate to lower the tip temperature to −70 °C (referred to as *sticking the probe*), which causes the probe to bond to the prostate tissue. Once all the probes are stuck in place, the actual freezing process begins, usually from anterior to posterior locations (Zippe 1996).

15.5.6 Research

Cryoablation has undergone major advances with the development of new imaging modalities such as intraoperative ultrasound, magnetic resonance imaging and infrared imaging in recent years. New applications for cardiac ablation have been widely investigated. Gallagher *et al* (1985) introduced cold mapping and cryoablation to refine the surgical treatment of accessory pathway-mediated tachycardia. They reported that cooling to 0 °C at the tricuspid annulus transiently abolished accessory pathway conduction and terminated tachycardia. Further cooling to −60 °C permanently interrupted accessory pathway conduction without the need for extensive surgical dissection.

Cold mapping and cryoablation have also been used to map and ablate perinodal atrial tissue participating in atrioventricular (AV) node re-entry. Dubuc *et al* (1998) tested a new steerable cryoablation catheter using Halocarbon 502 as a refrigerant. They found that reversible ice mapping of the AV node was successfully achieved in their animal studies. They suggested that this technology has the potential to allow for reversible ice mapping to confirm a successful ablation target before definitive ablation.

A novel cryothermia technique uses the Peltier effect to cool a semiconductor element at the tip of a catheter. The Peltier effect occurs when a voltage gradient is applied across two dissimilar materials in a thermocouple, resulting in cooling of one material and heating of the other. The material that is to be cooled is placed at the catheter tip. The heat produced in the second material is remote from the tissue interface and is removed by convection from the circulating blood pool. Such a system is likely to be dependent on adequate catheter–tissue contact in order to produce significant cooling of the target site. Holman and Rowland (1997) investigated the feasibility of using Peltier thermoelectric coolers (TECs) to cool down a suitable hand-held treatment tip to a temperature of approximately –50 °C. The new system does not use any disposable gases or liquid, and has a built-in temperature-sensing element at the tip. Clinical trials are being conducted to evaluate this new design.

15.6 MICROWAVE ABLATION

Microwave (MW) ablation produces lesions by tissue heating. Unlike RF ablation, MW electromagnetic energy can reach deeper tissue sites (potential advantage) directly, where it produces frictional heating by inducing dielectric ionic movements. Tissue heating in microwave ablation does not depend on the ability to pass electric current from the electrode, so it can penetrate through local tissue necrosis, old scars, or cartilaginous tissue. We discuss applications of MW ablation in this section.

15.6.1. Background

The quantification of microwave heating in flat layers of human tissues has been theoretically investigated by Schwan and Piersol (1954) for plane wave sources. Ho *et al* (1971) investigated MW heating for aperture sources. At frequencies above ~300 MHz, tissue acts as a lossy dielectric and the predominant mode of propagation for electromagnetic waves is radiative rather than conductive.

Some of the parameters that control the lesion sizes created by MW are (1) size of the antenna; (2) desired temperature levels; and (3) radiation patterns. Figure 15.16 shows the extent of Joule heating in RF ablation and the electromagnetic field induced by MW ablation. The electromagnetic field of MW ablation elevates tissue temperature both by conduction and displacement current. Thus, MW has the potential to heat a larger volume of tissue than RF ablation. MW ablation should also be less dependent on the catheter–tissue contact, and able to penetrate through desiccated tissue and blood clots and thus create a larger lesion.

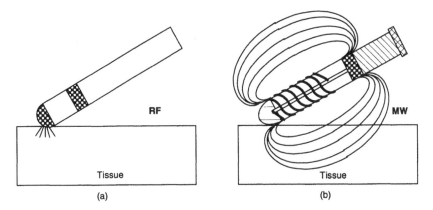

Figure 15.16. (*a*) Mechanism of tissue heating of RF ablation. (*b*) MW ablation produces an electromagnetic field and has a potential to create larger lesions. Adapted from Langberg and Leon (1995).

15.6.2 Designs

The MW ablation catheter system in figure 15.17 generally includes power supply, which is designed to generate controlled MW energy, a catheter, which is designed for insertion into the body of a patient, and a connector for coupling the power supply to the catheter. The power supply includes a casing, an oscillator, a waveguide adapter, a pair of directional couplers that interface with power monitors, a tuner, a controller and an interlock system all enclosed. The front panel of the casing has various displays and controls.

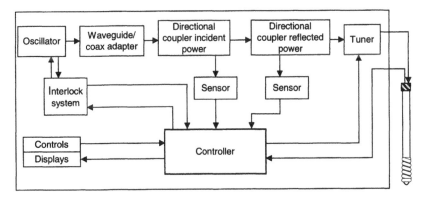

Figure 15.17. A schematic block diagram of a MW power supply system for an ablation catheter. Adapted from Warner and Grundy (1994).

The MW generator may take any conventional form. When using MW energy for tissue ablation, the optimal frequencies are generally in the neighborhood of the optimal frequency for heating water. The frequencies that are approved by the US Food and Drug Administration for experimental clinical use are 915 MHz and 2.45 GHz. Therefore, a power supply having the capacity to generate MW energy at frequencies in the neighborhood of 2.45 GHz was chosen.

The best small diameter waveguides for transmitting MW energy are transmission lines that take the form of coaxial cables. Thus, a conventional waveguide adapter couples the MW generator to a coaxial cable. A pair of directional couplers are provided downstream of the waveguide adapter. The output of each is coupled to a power sensor whose signal is processed by the controller. Conventional power sensors may be used. The purpose of the directional coupler/power sensor arrangements is to monitor the power outputted by the generator as well as the reflected power.

15.6.3 Advantages and limitations

MW ablation has an advantage over other modalities in its titratable property. Tissue heating starts slowly with MW ablations, but continues to occur beyond 60 s, resulting in further lesion growth, thus allowing better control of lesion size. MW ablation may also avoid some of the complications associated with energy delivery by RF conduction currents, such as the formation of a coagulum over the catheter tip at high temperatures, which increases the electrode impedance and severely impairs energy delivery.

A potential disadvantage of the MW technique is dissipation of MW power along the transmission line, due to dielectric losses and reflection back from antenna. This can result in heating of the catheter shaft with the potential for thermal injury at sites remote from the tip.

15.6.4 Applications

15.6.4.1 MW catheter ablation for the treatment of arrhythmias. Recently, current in the MW range has been explored as an energy source for cardiac ablation. RF lesion size is critically dependent on the surface area of the electrode–tissue interface and on the conduction of heat from the point of contact into the adjacent myocardium. In contrast, an MW antenna radiates an electromagnetic field into the surrounding tissue. While maintaining alignment with the alternating electric field, neighboring water molecules collide with each other and generate heat primarily as the result of friction. Based on this mechanism of action, MW energy has the potential to heat a larger volume of myocardium than can be ablated using a conventional radio-frequency catheter and should be less dependent on catheter contact area or contact pressure.

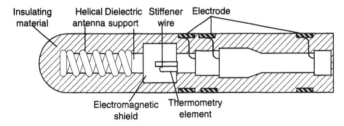

Insulating material   Helical   Dielectric   Stiffener   Electrode
antenna support   wire

Electromagnetic   Thermometry
shield   element

Figure 15.18. Diagram of helical antenna MW electrode for cardiac ablation. The stiffener wire allows better flexure control of the catheter. The electromagnetic shield prevents the intense field in the middle of the coil from the wire and electrodes. The insulating material (e.g. Teflon) helps avoid charring and coagulation. Adapted from Warner and Grundy (1994).

Figure 15.18 shows an example of a helical antenna catheter. If good matching with tissue permittivity could be accomplished, MW energy can produce a large volume of heating without causing endocardial charring from overheating. An insulator like a heat shrink tubing is placed over the applicator to enhance impedance matching.

Other types of MW applicator are (Shetty *et al* 1996):

1. I-radiators;
2. U-radiators;
3. O-radiators;
4. forward helical coil radiator;
5. reverse helical coil radiator;
6. double coil radiator;
7. loaded monopole radiator;
8. leaky coaxial radiator;
9. tee radiators.

15.6.4.2 MW surgical treatment of the prostate. Open surgery or transurethral resection (TUR) has been used to relieve bladder neck obstruction resulting from prostatic enlargement. Although these procedures are well accepted, they are also occasionally complicated by severe hemorrhage and are also associated with other complications especially in elderly or high-risk patients. Cryoprostatectomy has been advocated as an alternative treatment in the high-risk patient and in patients with potential bleeding diathesis.

MW prostatic ablation can cause significant destruction of prostatic tissue and also relieve the obstruction. The procedure can be used simply and safely because the depth of coagulation is controlled by the length of the radiating electrodes of the probe tip, the electric output intensity and the duration of the MW exposure. However, therapy by MW prostatic coagulation alone has the problem of persistent sloughing of necrotic tissue following the procedure.

The advantage of this method found in clinical trials is the absence of hemorrhage both during and after the procedure. However, similar problems are encountered as in cryoprostatectomy. For example, prolonged catheterization is more frequently required than with ordinary TUR, because the necrotic tissue takes time to fall off. To avoid such problems, Harada *et al* (1987) performed MW prostatic coagulation followed by immediate TUR.

15.6.5 Research

The majority of recent studies on MW ablation have been focused on applications for cardiac catheter ablation. Whayne *et al* (1994) investigated monopole antennas operating at 915 and 2450 MHz for MW ablation of canine myocardium *in vitro*, and found that MW lesion depths were indeed larger than those produced by RF energy. Rosenbaum *et al* (1993) also investigated helical and whip antennas with *in vivo* and phantom models in order to optimize the operating frequency and antenna design. Kaouk *et al* (1996) also illustrated that the temperature distributions produced by MW energy have a larger extent than those produced by RF energy delivered by the same catheter. Thomas *et al* (1999) investigated the effects of blood flow, power and duration of exposure on lesion dimensions.

Haugh *et al* (1997) examined the ability of MW pulsing to achieve a more uniform myocardial temperature gradient in an *in vitro* setting. They delivered MW energy at 20 W at 915 MHz, and compared pulse configurations of 1 s on–1 s off, 3 s on–3 s off, and 5 s on–5 s off, with 30 s of total MW time, with 30 s continuous energy delivery. They monitored the temperatures in tissue and discovered that lower tissue temperatures were found at 0.5 mm from the tissue surface for the 1 s pulse case, compared with the continuous-energy delivery. Pulse configurations 3 s on–3 s off and 5 s on–5 s off also resulted in a significantly lower surface temperature than continuous-energy delivery. However, temperature at the 2.0 mm and 3.5 mm depth created by the pulsing delivery were similar to those achieved during continuous-energy delivery. They concluded that MW pulsing achieves a lower endocardial temperature and results in a more uniform temperature gradient. These techniques may prevent the excessive endocardial damage that may result in an increased risk of thrombus formation and embolization.

15.7 CHEMICAL ABLATION

Chemical substances can also be used to ablate unwanted tissues. As early as 1964 a complete heart block was induced in dogs by the injection of formaldehyde in the region of the bundle of His. Damiano *et al* (1986) could lower the fibrillation threshold of ventricular myocardium in dogs by administering intraoperatively Lugol's solution into the ventricular endocardium. A suitable chemical for ablation must be safe and effective and require a brief contact time.

15.7.1 Applications of chemical ablation

15.7.1.1 Chemical ablation of the gallbladder. Gallstones can be removed by instillation of chemicals into the lumen of the gallbladder. One of the advantages of this technique is that there is no need for general anesthesia (only regional anesthesia). A procedure for chemical ablation of the bladder includes a midline laparatomy, an introduction of intravenous cannula into the gallbladder. The gallbladder is then flushed with a chemical (see table 15.2). The chemical is left in contact for approximately 5–30 min.

Majeed *et al* (1997) summarizes the advantages and disadvantages of several chemicals that are already in clinical use. Table 15.2 lists the mechanism of action, advantages and disadvantages of each chemical.

Table 15.2. Summary of advantages and disadvantages of various chemicals for gallbladder chemical ablation. From Majeed *et al* (1997).

Chemical	Mechanism of action	Advantages	Disadvantages
Heated contrast medium	Thermal destruction	Visible on fluoroscopy, safe	Not effective
Cyanoacrylate nitrocellulose	Adhesion to tissue	Possible elimination of the need to occlude cystic duct if glue is viscid	Not effective for mucosal ablation
Absolute alcohol and 95% ethanol	Dehydration and denaturation	Safe	Partially effective; prolonged contact may lead to significant absorption
Sodium carbonate	Direct toxicity	Effective	Toxic if leaks out of gallbladder
Tetracycline	Direct toxicity	Safe	Effective only after prolonged contact
Sodium tetradecyl sulfate	Direct toxicity	Safe	Not effective when used alone

15.7.1.2 Chemical ablation for ventricular tachycardia. Several recent studies on chemical ablation for cure of cardiac arrhythmias have been presented. Verna *et al* (1992) reported that transmural myocardial lesion, necessary in the ablation of ventricular tachycardia associated with coronary artery disease, could be created by infusing 5 ml or higher of ethanol. However, some complications, such as (1) short-term chest pain and (2) complete heart block due to occlusion of septal artery, can occur during this procedure. Weismuller *et al* (1991) performed a direct subendocardial injection of ethanol via a catheter *in vivo* and were able to create lesion volumes of approximately 60 mm^3, depths of 1–5 mm, and maximum diameters of 8 mm. Wright *et al* (1998) reported that infusions of ethanol using a balloon-tipped infusion catheter were effective in ablating ventricular myocardium in clinical studies.

PROBLEMS

15.1 Explain the advantages and disadvantages of RF ablation.

15.2 Describe the bio-heat equation.

15.3 Explain why a thermistor is embedded at the tip of the RF electrode.

15.4 Explain absorption, spontaneous emission and stimulated emission.

15.5 Explain the differences between CO_2, Nd:YAG and argon lasers.

15.6 Explain the advantages and disadvantages of laser ablation.

15.7 Describe how the laser system for removing cataracted lens tissue works.

15.8 Draw a diagram of ultrasound ablation and explain how it works.

15.9 Describe the procedure for prostate cryoablation.

REFERENCES

Bromer R H, Mitchell J B and Soares N 1982 Response of human hematopoietic precursor cells (CFUc) to hyperthermia and radiation *Cancer Research* **42** 1261–5

Brucker G G, Saul J P and Savage S D 1997 *Fluid cooled and perfused tip for a catheter* US patent 5,643,197

Cain C A, Ebbini E S and Strickberger S A 1997 *Phased array ultrasound system and method for cardiac ablation* US patent 5,590,657

Damiano R J Jr, Smith P K, Tripp H F Jr, Asano T, Small K W, Lowe J E, Ideker R E and Cox J L 1986 The effect of chemical ablation of the endocardium on ventricular fibrillation threshold *Circulation* **74** 645–52

Dubuc M, Talajic M, Roy D, Thibault B, Leung T K and Friedman P L 1998 Feasibility of cardiac cryoablation using a transvenous steerable electrode catheter *J. Interventional Cardiac Electrophysiol.* **2** 285–92

Duck F A, Baker A C and Starritt H C 1998 *Ultrasound in Medicine* (Bristol: Institute of Physics)

Edwards S D and Stern R A 1997 *Electrode and associated systems using thermally insulated temperature sensing elements* US patent 5,688,266

Foster R S, Bihrle R, Sanghvi N T, Fry F J and Donohue J P 1993 High-intensity focused ultrasound in the treatment of prostatic disease *Eur. Urol.* **23** Suppl 1 29–33

Fry F J 1993 Intense focused ultrasound in medicine: some practical guiding physical principles from sound source to focal site in tissue *Eur. Urol.* **23** Suppl 1 2–7

Fry W J and Fry F J 1960 Fundamental neurological research and human neurosurgery using intense ultrasound *IRE Trans. Med. Electron.* **7** 166–81

Fry W J, Mosberg W H, Barnard J W and Fry F J 1954 Production of focal destructive lesions in the central nervous system with ultrasound *J. Neurosurg. Psychiatry* **11** 471–8

Gallagher J D, Del Rossi A J, Fernandez J, Maranhao V, Strong M D, White M and Gessman L J 1985 Cryothermal mapping of recurrent ventricular tachycardia in man *Circulation* **71** 732–9

Gelet A, Chaperlon J Y, Margonari J, Theilliere Y, Gorry F, Souchon R and Bouvier R 1993 High-intensity focused ultrasound experimentation on human benign prostatic hypertrophy *Eur. Urol.* **23** Suppl 1 44–7

Goldberg S N and Gazelle G S 1998 Advances in radiofrequency tumor ablation therapy: technical considerations, strategies for increasing coagulation necrosis volume, and preliminary clinical results *Proc. Surg. Appl. Energy SPIE* **3249** 104–14

Goldberg S N, Gazelle G S, Solbiati L, Rittman W J and Mueller P R 1996 Radiofrequency tissue ablation: increased lesion diameter with a perfusion electrode *Acad. Radiol.* **3** 636–44

Goldberg S N, Solbiati L, Hahn P F, Cosman E, Conrad J E, Fogle R and Gazelle G S 1998 Large-volume tissue ablation with radio frequency by using a clustered, internally cooled electrode technique: laboratory and clinical experience in liver metastases *Radiol.* **209** 371–9

Haar G R and Robertson D 1993 Tissue destruction with focused ultrasound *in vivo Eur. Urol.* **23** Suppl 1 8–11

Haines D E 1995 The biophysics and pathophysiology of lesion formation during radiofrequency ablation *Cardiac Electrophysiology: from Cell to Bedside* 2nd edn, ed D P Zipes (Philadelphia, PA: Saunders)

Harada T, Tsuchida S, Nishizawa O, Kigure T, Noto H, Etori K, Kumazaki T, Koh D and Shimoda J 1987 Microwave surgical treatment of the prostate: clinical application of microwave surgery as a tool for improved prostatic electroresection *Urol. Int.* **42** 127–31

Haugh C, Davidson E S, Estes N A III and Wang P J 1997 Pulsing microwave energy: a method to create more uniform myocardial temperature gradients. *J. Interventional Cardiac Electrophysiol.* **1** 57–65

Ho H S, Guy A W, Lehmann J F and Sigelmann R A 1971 Microwave heating of simulated human limbs by aperture sources *IEEE Trans. Microw. Theory Tech.* **19** 224–31

Holman M R and Rowland S J 1997 Design and development of a new cryosurgical instrument utilizing the Peltier thermoelectric effect *J. Med. Eng. Technol.* **21** 106–10

Huang S K S and Wilber D J (eds) 2000 *Radiofrequency Catheter Ablation of Cardiac Arrhythmias: Basic Concepts and Clinical Applications* 2nd edn (Armonk, NY: Futura)

Jeffreys R V 1992 *Lasers in Neurosurgery* (New York: Wiley–Liss)

Kaouk Z, Khebir A and Savard P 1996 A finite element model of a microwave catheter for cardiac ablation *IEEE Trans. Microw. Theory Tech.* **44** 1848–54

L'Esperance F A Jr 1985 *Method and apparatus for removing cataractous lens tissue by laser radiation* US patent 4,538,608

Langberg J J and Leon A 1995 Energy sources for catheter ablation *Cardiac Electrophysiology: from Cell to Bedside* 2nd edn, ed D P Zipes (Philadelphia, PA: Saunders)

Lorentzen T, Christensen N E, Nolsle C P and Torp-Pedersen S T 1997 Radiofrequency tissue ablation with a cooled needle *in vitro*: ultrasonography, dose response, and lesion temperature *Acad. Radiol.* **4** 292–7

Lynn J G, Zwemer R L, Chick A J and Miller A E 1942 A new method for the generation and use of focused ultrasound in experimental biology *J. Gen. Physiol.* **26** 179–92

Majeed A W, Reed, M W R, Stephenson T J and Johnson A G 1997 Chemical ablation of the gallbladder *Br. J. Surg.* **84** 638–41

McGahan J P, Gu W Z, Brock J M, Tesluk H and Jones C D 1996 Hepatic ablation using bipolar radiofrequency electrocautery *Acad. Radiol.* **3** 418–22

McRury I D, Panescu D, Mitchell M A and Haines D E 1997 Nonuniform heating during radiofrequency catheter ablation with long electrodes: monitoring the edge effect *Circulation* **96** 4057–64

Nath S, Lynch C III, Whayne J G and Haines D E 1993 Cellular electrophysiological effects of hyperthermia on isolated guinea pig papillary muscle: implications for catheter ablation *Circulation* **88** Part 1 1826–31

Onik G, Cobb C, Cohen J, Zabkar J and Porteffield B 1988 US characteristics of frozen prostate *Radiol.* **168** 629–31

Panescu D 1997 Intraventricular electrogram mapping and radiofrequency cardiac ablation for ventricular tachycardia *Physiol. Meas.* **18** 1–38

Panescu D, Whayne J G, Fleischman D, Mirotznik M S, Swanson D K and Webster J G 1995 Three-dimensional finite element analysis of current density and temperature distributions during radio-frequency ablation *IEEE. Trans. Biomed. Eng.* **42** 870–90

Parrish J A and Wilson B C 1991 Current and future trends in laser medicine *Photochem. Photobiol.* **53** 731–8

Patel C K N, McFarlane R A and Faust W L 1964 Selective excitation through vibrational energy transfer and optical laser action in N_2–CO_2. *Physiol. Rev.* **4** 182

Rosenbaum R M, Greenspon A J, Hsu S, Walinsky P and Rosen A 1993 RF and microwave ablation for the treatment of ventricular tachycardia *1993 IEEE MTT-S Int. Microwave Symp. Digest* **2** 1155–8

Rossi S, Di Stasi M, Buscarini E, Quaretti P, Garbagnati F, Squassante L, Paties C T, Silverman D E and Buscarini L 1996 Percutaneous RF interstitial thermal ablation in the treatment of hepatic cancer *Am. J. Roentgenol.* **167** 759–68

Schulman C C and Vanden Bossche M 1993 Hyperthermia and thermotherapy of benign prostatic hyperplasia: a critical review *Eur. Urol.* **23** Suppl 1 53–9

Schwan H P and Piersol G M 1954 The absorption of electromagnetic energy in body tissues *Am. J. Phys. Med.* **33** 370–404

Shetty S, Ishii T K, Krum D P, Hare J, Mughal K, Akhtar M and Jazeyeri M R 1996 Microwave applicator design for cardiac tissue ablations *J. Microw. Power Electromagn. Energy* **31** (1) 59–66.

Silverman R H, Vogelsang B, Rondeau M J and Coleman D J 1991 Therapeutic ultrasound for the treatment of glaucoma *Am J. Ophthalmol.* **111** 887–91

Sliney D H and Trokel S L 1993 *Medical Lasers and their Safe Use* (New York: Springer-Verlag)

Svenson R H, Selle J G, Gallagher J J, Marroum M-C, Tatsis G P and Seifert K T 1987 Neodymium: YAG laser, photocoagulation: a potentially useful method for intraoperative ablation of arrhythmogenic foci *Ablation in Cardiac Arrhythmias* ed G Fontaine and M M Scheinman (Mount Kisco, NY: Futura)

Thomas S P, Clout R, Deery C, Mohan A S and Ross D L 1999 Microwave ablation of myocardial tissue: the effect of element design, tissue coupling, blood flow, power, and duration of exposure on lesion size *J. Cardiovasc. Electrophysiol.* **10** 72–8

Tungjitkusolmun S, Woo E J, Cao H, Tsai J-Z, Vorperian V R and Webster J G 2000 Finite element analyses of uniform current density electrodes for radio-frequency cardiac ablation *IEEE Trans. Biomed. Eng.* **47** 32–40

Verdier J 1992 The cryogens *Cryotherapy in Chest Medicine* ed J P Homasson and N J Bell (New York: Springer)

Verna E, Repetto S, Saveri C, Forgione N, Merchant S and Binaghi G 1992 Myocardial dissection following successful chemical ablation of ventricular tachycardia *Eur. Heart J.* **13** 844–6

Warner G G and Grundy D A 1994 *Microwave ablation catheter system with impedance matching tuner and method* US patent 5,364,392

Weber H, Enders S and Keiditisch E 1989 Percutaneous Nd:YAG laser coagulation of ventricular myocardium in dogs using a special electrode laser catheter *PACE* **12** 899–910

Weismuller P, Mayer U, Richter P, Heieck F, Kochs M and Hombach V 1991 Chemical ablation by subendocardial injection of ethanol via catheter—preliminary results in the pig heart *Eur. Heart J.* **12** 1234–9

Welch A J, Motamedi M, Rastegar S, LeCarpentier G L and Jansen D 1991 Review article: laser thermal ablation *Photochem. Photobiol.* **53** 815–23

Whayne J G, Nath S and Haines D E 1994 Microwave catheter ablation of myocardium *in vitro*. Assessment of the characteristics of tissue heating and injury *Circulation* **89** 2390–95

Wright K N, Morley T, Bicknell J, Bishop S P, Walcott G P and Kay G N 1998 Retrograde coronary venous infusion of ethanol for ablation of canine ventricular myocardium *J. Cardiovasc. Electrophysiol.* **9** 976–84

Yang R, Sanghvi N T, Rescorla F J, Kopecky K K and Gorsfeld J L 1993 Liver cancer ablation with extracorporeal high-intensity focused ultrasound *Eur. Urol.* **23** Suppl 1 17–22

Zimmer J E, Hynynen K, He D S and Marcus F 1995 The feasibility of using ultrasound for cardiac ablation *IEEE Trans. Biomed. Eng.* **42** 891–7

Zipes D P 1994 *Catheter Ablation of Arrhythmias* (Armonk, NY: Futura)

Zippe C D 1996 Cryosurgery of the prostate: techniques and pitfalls *Urol. Clin. N. Am.* **23** 147–63

CHAPTER 16

NEUROMUSCULAR STIMULATION

John G Webster

The first neuromuscular stimulation was external. Nerves in the brain were stimulated by electroshock. With modern technology, stimulators have been made small enough to be implanted within the brain to control Parkinson's disease. Patients who had lost the ability to contract their muscles, have their muscles stimulated by surface electrodes on the belly of the muscle. Stimulators have also been made small enough to be implanted next to the motor nerves that feed the muscles. Other small implants stimulate the phrenic nerve to assist breathing, the nerves that control voiding and cardiac pacemakers and implantable cardioverter–defibrillators to provide therapy for the heart.

16.1 STIMULATING NERVE

The resting potential of the interior of excitable nerve and muscle membranes is about -70 mV. When the current density from a stimulator is large enough to cause the membrane potential to become more positive by about 20 mV, this causes an action potential to occur. The action potential propagates down the nerve. At the end of the nerve, through synapses it may excite other nerves. When motor nerves are excited, they feed muscles and cause muscle contraction. A less efficient way to stimulate muscle is through electrodes on the surface of the muscle.

16.1.1 Brain stimulation

Parkinson's disease causes muscular tremor. Patients may have an electric stimulator implanted in the brain. The patient can control the level of stimulation to achieve the best level of therapy. Control signals are transmitted by radio waves from the patient's keypad to a receiver in the stimulator. A similar system is available for controlling rage.

16.1.2 Diaphragm stimulation

Patients whose diaphragm function is impaired may have it contracted by small cuff electrodes placed around the phrenic nerve. The level of stimulation increases with a ramp function, thus causing a ramp-like function in diaphragm contraction and lung inspiration (Dobelle Group 2000).

16.1.3 Bladder stimulation

Patients with problems voiding the bladder may have cuff electrodes placed around the nerves leading to the detrusor muscle that assists bladder voiding and also to the sphincter muscle that must relax for voiding.

16.2 CARDIAC PACEMAKERS

When the atrioventricular node fails to conduct, heart block occurs because excitation in the atria fails to conduct to the ventricles. It is possible to pace the heart using transcutaneous pacing by passing 100 mA, 2 ms pulses through large electrodes on the chest, but undesired stimulation of the intercostal muscles causes great pain. The advent of miniature implantable pacemakers has revolutionized the treatment of heart block and many other cardiac rhythm disorders (Webster 1995).

16.2.1 Lead

Under local anesthesia, a 10 cm incision is made beneath the left collarbone to place the transvenous lead. The biocompatible polyurethane or silicone rubber insulated flexible wire lead is threaded through an incision in the vein leading to the heart, through the right atrium, through the tricuspid valve, and the tip electrode embedded in the apex of the right ventricle. The 4 mm^2 electrode tip is of porous platinum–iridium and may include steroids that are eluted to minimize inflammation.

16.2.2 Power source

About half of the pacemaker volume is taken up by the battery, which lasts about a decade. The usual battery is lithium–iodide, which yields no gas and is hermetically sealed. It provides 3 A h charge at 2.8 V. Near the end of life, the voltage drops, a circuit senses it, and changes the heart rate to provide an indication that the pacemaker should be changed.

16.2.3 Sensing

The lead senses electrograms from the myocardium. A sense amplifier amplifies these so the computer can determine if the rate is satisfactory. If so, the pacemaker does not pace. If the heart rhythm is incorrect, the pacemaker provides pulses at the proper rate.

16.2.4 Control

The control unit contains several timers. If the heart rate is normal, each sensed beat resets the timers. If the heart rate is too slow, a timer times out and the pacemaker paces the heart. Other timers control a blanking period that prevents the pacemaker from pacing too fast. The control unit may also store selected waveforms and other data for playback when interrogated.

16.2.5 Pulse-generating unit

The pulse generator generates pulses of sufficient intensity so that the electrode stimulates the myocardium. It contains an output capacitor to block dc that would corrode the electrode and cause tissue inflammation. It may contain a voltage doubling circuit to boost the voltage to 5.4 V. Stimulation is varied by varying the pulse width from 1 to 2 ms.

16.2.6 Pacing synchrony

In asynchronous (fixed rate) pacing, the heart chamber is paced at a constant rate regardless of the physiological needs of the body. If the ventricles are paced, they are not synchronous with the atria.

In synchronous (demand) pacing, the heart is paced based on an event in the heart. If the heart has a natural beat, this inhibits the pacemaker from supplying a pulse. If the heart does not supply a natural beat after a preset duration, the pacemaker supplies a pulse (on demand).

In rate-adaptive pacing, a sensor, such as an accelerometer, senses body movement (such as walking) and increases the heart rate to supply the increased cardiac output required to sustain the movement (Schaldach 1992).

16.3 IMPLANTABLE CARDIOVERTER–DEFIBRILLATORS

If the heart goes into sustained rapid rate (tachycardia), this dysrhythmia may be terminated by a large synchronous pulse. This is called cardioversion. If the heart goes into chaotic beating (fibrillation), this dysrhythmia may be terminated by a larger asynchronous pulse so that the entire myocardium becomes refractory. This is called defibrillation. Modern technology has enabled the implantable cardioverter–defibrillator (ICD) to be small enough to be implanted in a pocket under the collarbone and the pulses applied through a transvenous lead. This reduces the incidence of sudden cardiac death (SCD), which is death that occurs within 24 h of onset of symptoms.

PROBLEMS

16.1 Sketch the cell membrane potential versus time for a slowly increasing stimulation current density.

16.2 Sketch the amplitude versus pulse width to achieve equal stimulation of nerve. Repeat for muscle (Geddes and Baker 1989).

16.3 Sketch the placement of a passive implantable nerve stimulation system (Dobelle 2000).

16.4 Describe the cardiac conditions that suggest implantation of a cardiac pacemaker (Webster 1995).

16.5 Describe the stimulation threshold versus time after cardiac pacemaker implantation (Webster 1995).

16.6 Sketch the block diagram for a demand pacemaker (Webster 1995).

16.7 Sketch the block diagram for a rate-adaptive cardiac pacemaker (Schaldach 1992).

16.8 Sketch the output circuit for a cardiac pacemaker (Webster 1995).

16.9 Describe the cardiac conditions that suggest implantation of a cardioverter–defibrillator (Webster 1995).

16.10 Explain how the battery for a cardioverter–defibrillator differs from that of a cardiac pacemaker (Webster 1995).

16.11 Explain cardiac tiered therapy (Webster 1995).

16.12 Sketch the current pathways for cardioverter–defibrillator patch electrodes and transvenous electrodes (Webster 1995).

REFERENCES

Dobelle Group 2000 website http://www.dobelle.com/
Geddes L A and Baker L E 1989 *Principles of Applied Biomedical Instrumentation* 3rd edn New York: John Wiley), chapter 10
Schaldach M M 1992 *Electrotherapy of the Heart* (Berlin: Springer)
Webster J G (ed) 1995 *Design of Cardiac Pacemakers* (Piscataway, NJ: IEEE)

CHAPTER 17

HELICAL TOMOTHERAPY

G H Olivera, J M Kapatoes, K J Ruchala,
P J Reckwerdt, E E Fitchard, H Keller,
J P Balog, T R Mackie

17.1 INTRODUCTION

In radiotherapy high-energy beams such as photons, electrons, protons, neutrons, heavy ions, etc are used as a means to treat cancer. The objective of radiotherapy is to provide tumor control while compromising as little as possible the life quality of the patient. The vast majority of radiotherapy treatments were performed with photons. For a long period of time, ^{60}Co was the main radioactive source used for cancer treatment. During the 1970s the linear accelerators (linacs) gradually replace the cobalt machines. During the late 1980s independent investigations developed by Cormack and Brahme (Brahme 1988, Cormack 1987, Cormack and Cormack 1987) showed that treatments could be drastically improved by using beams with modulation of their intensity. The use of therapeutic beams with modulation of the intensity within the field size is usually referred to as intensity modulated radiotherapy (IMRT). The use of modulated beams is not new in radiotherapy. Devices such as wedges were used long ago to provide beams of variable intensity. However, the advance in technology provides the possibility to produce more sophisticated modulation devices such as multi-leaf collimators (MLCs). The more commonly used of such devices are dynamic (Bortfeld *et al* 1994, Mohan *et al* 1996, Purdy 1996, van Santvoort and Heijmen 1996, Verhey 1996, Webb 1997) and temporal (Carol *et al* 1992, Sternick 1997, Swerdloff *et al* 1994, Webb 1997) or binary MLCs. Figure 17.1(*a*) shows a schematic representation of the temporal MLC. A set of 64 tungsten leaves move in or out of the radiation field regulating the time and location that the radiation is delivered to the patient. Figure 17.1(*b*) is a picture of the MLC under development at the University of Wisconsin–Madison.

a   b

Figure 17.1. (*a*) Schematic representation of the temporal MLC. (*b*) Picture of the MLC under development at the University of Wisconsin–Madison.

IMRT provides the possibility to perform conformal and conformal avoidance radiotherapy. Conformal radiotherapy (CR) is aimed to tightly conform the dose distribution to the tumor, sparing as much as possible healthy tissue. CR is well suited to treat patients where the tumor boundary can be easily identified. Sometimes, however, it is very difficult to know the actual extent of the tumor, therefore, it is not rational to conform the dose to the tumor. On the contrary, the regions that should not be irradiated (avoidance regions) in general may be easy to identify in computed tomography (CT) or in magnetic resonance images (MRI). Conformal avoidance (CA) is a valid alternative in this case, because the irradiation can be performed over a region where the tumor is suspected, while maintaining the regions at risk below allowed levels.

Several methodologies of IMRT are available (Bortfeld *et al* 1994, Carol *et al* 1992, Gustafsson *et al* 1994, Ling *et al* 1996, Low and Mutic 1998, Oldham and Webb 1997, Oldham *et al* 1998, Spirou and Chui 1994, Sternick 1997, Svensson *et al* 1994, van Santvoort and Heijmen 1996, Webb 1991, 1997, Xia and Verhey 1998, Yu *et al* 1995). All of these techniques when used in conformal or conformal avoidance radiotherapy have in common the presence of high dose gradients in the tumor region and/or the sensitive structures. Therefore, in CR and CA radiotherapy, it is not only important to have IMRT capabilities but also sophisticated verification capabilities necessary to guarantee that the dose is adequately delivered to the patient.

During the early 1990s Mackie and co-workers proposed the tomotherapy concept (Mackie 1997, Mackie *et al* 1993, 1997, 1999, Olivera *et al* 1999). The first clinical prototype of this machine is under development at the University of Wisconsin–Madison. Tomotherapy literally means slice therapy. Figure 17.2 shows an artist's rendering of the machine. Basically a linac with beam modulation capabilities is mounted in a CT-like ring gantry. The linac and MLC rotate around the patient as the couch translates, producing a helical delivery pattern. This helical dose delivery reduces the dose junctioning problems (Balog 1996, 1998, Balog *et al* 1999a, b, Carol *et al* 1996, Low 1997) presented in another approach called sequential tomotherapy, which is being pursued by NOMOS Corporation (Carol *et al* 1992). In sequential tomotherapy, the patient is indexed after each rotation is completed. Helical tomotherapy also includes a megavoltage detector opposite to the linac that provides tomographic capabilities. Moreover, if a kilovoltage tube and the corresponding CT detector are mounted, the unit could also provide the capability for diagnostic-quality CT images.

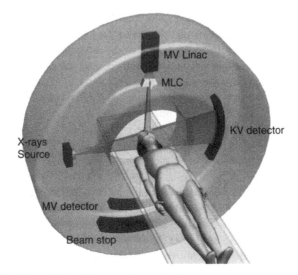

Figure 17.2. Artist's rendering of the tomotherapy machine.

Such a combination of several devices into a single piece of equipment provides the possibility to perform processes that are very difficult or even impossible for other IMRT modalities. This hardware, in conjunction with the set of processes to be described, makes helical tomotherapy a closed loop for planning, delivery and verification in radiotherapy.

17.2 PROCESSES

Figure 17.3 is a flow chart of the typical tomotherapy process. First, the patient will have a set of CT images to perform three-dimensional (3D) treatment planning. With these CT images the physician will prescribe a radiotherapy treatment specifying the location of tumors and involved nodal regions that should be irradiated, and the regions that must be avoided and/or the dose minimized. Because a typical tomotherapy treatment will involve hundreds of thousands of 'beam lets' and millions of volume elements describing the patient's anatomy, a computer program is used to calculate the intensity of the beam lets that best implement the prescription of the physician and physicist. This process, called treatment optimization, is the next step in the tomotherapy sequence. Before the optimized treatment plan can be delivered, the patient must be positioned accurately. Compared to standard radiotherapy, a greater delivery accuracy is required due to the presence of high dose gradient regions near to the tumor regions and/or sensitive structures. In tomotherapy such accuracy can be achieved in two different ways. One is to utilize the CT imaging capabilities of the machine. This type of localization may be preferred for the first treatment in a course of therapy. However, the image reconstruction and its 3D registration process usually is time consuming; however, a tomotherapy process called patient registration in projection space can accomplish the positioning registration of the patient in a few seconds and reduces the dose deposited to the patient. During the first treatment a loose (pitch greater than one) helical scan of the patient in the correct position is obtained. Pitch is defined as the ratio between the couch translation per linac rotation and the slice width in the direction of translation. During this scan the signal collected by the exit detector is recorded as a 2D array called a registration sinogram. The abscissa of this sinogram is the detector number and the ordinate is the projection number. The sinogram contains the same information as the CT does about the patient's position. The registration sinogram obtained in the first treatment will be the reference position for some of the subsequent treatments in the course of therapy.

In all the other fractions another loose helical sinogram is obtained and by comparison with the reference one it is possible to obtain very accurate information about the possible three translational and three rotational displacements of the patient. The convenience of the registration in projection space is that time-consuming image reconstruction is avoided.

Once the displacement of the patient is known, it is possible to reposition the patient, which may be an error-prone technique, or the treatment delivery can be modified. Delivery modification is a process that uses the information retrieved by the patient registration algorithm, modifying in a few seconds the delivery sinogram, adapting it to the movement of the patient. In delivery modification there is no need to re-optimize the treatment plan, which may be very time consuming.

After these processes are accomplished the treatment can be delivered. During delivery two other processes can be started. One is the radiation delivery verification process, which uses the signal measured at the exit detector to compute the energy fluence actually delivered by the linac on a pulse-by-pulse basis. Any error during delivery out of the tolerance will shut down the machine, recording the information of the status of the treatment at that time. The other process that can be done during delivery is the acquisition of kV or MV CT.

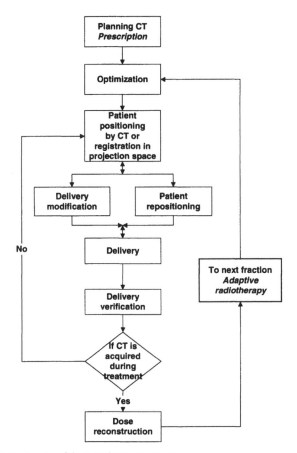

Figure 17.3. Block diagram of the tomotherapy processes.

By combining the information obtained during the radiation delivery verification process and the CT image before or during treatment, dose reconstruction can be performed. This technique provides the dose actually deposited in the patient. Comparison between planned and actually delivered treatment can be done. The isodose lines of the treatment actually delivered can be plotted over the CT obtained during that particular delivery. Using dose reconstruction it is possible to perform an adaptive radiotherapy whereby the plan for subsequent treatments is modified to take into account deviations in the treatments to date.

17.2.1 Optimization

After a plan is prescribed, a computer program is used to compute the delivery sinogram to best deliver the prescription. This sinogram is represented by a 2D array and contains the information about where and how long every leaf must be open. The columns represent leaf number and the rows represent the projection number. As in any optimization an objective function is used to score the plan together with constraints and/or penalties. The objective function can be based on physical and/or biological parameters. Due to the lack of complete information on which to base biological models, at this stage, tomotherapy optimization is based on physical objective functions.

An example of an optimization for a prostate treatment will be presented. In this case the objective function used is the quadratic objective function shown in equation (17.1):

$$O(w) = \sum_{i \in T} C_T \left(d_i^{\mathrm{p}} - d_i^{\mathrm{d}} \right)^2 + \sum_{i \in \mathrm{RAR}} C_{\mathrm{RAR}} \, \lambda_{\mathrm{DVH}} \left(d_i^{\mathrm{p}} - d_i^{\mathrm{d}} \right)^2 \qquad (17.1)$$

where w is a vector representing the incident energy fluence, T and RAR represent tumor and region-at-risk voxels, respectively, d_i^{p} is the dose prescribed to the voxel i, d_i^{d} is the dose delivered to the voxel i, C_T and C_{RAR} are the importance weights assigned to the tumor and regions at risk, respectively, and λ_{DVH} is the weight of the dose volume histogram (DVH) penalty. A more detailed discussion about the implementation of the optimization can be found elsewhere (Olivera *et al* 1999, Shepard *et al* 1999, 2000). In the case presented, the prescription to the prostate was that more than 90% of the volume receives more than 75 Gy. Moreover, a DVH penalty was applied to the bladder and the rectum if more than 20% of the volume received more than 30 Gy.

Figures 17.4(*a*) and (*b*) are the isodose distributions for one of the slices of the prostate superimposed with the RAR and the CT, respectively. It is possible to observe that the isodoses tightly conform to the tumor and spare adequately the regions at risk. A more quantitative analysis of the treatment can be seen in the DVH presented in figure 17.5. The DVH plot is used to analyze the dose received by each volume fraction of the regions of interest (ROIs).

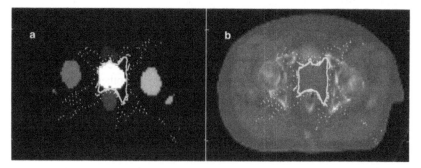

Figure 17.4. (*a*) ROIs for the prostate treatment with isodose lines: dotted line 50%, dashed line 80%, full line 95% isodose lines; (*b*) CT for the prostate treatment with isodose lines: dotted line 50%, dashed line 80%, full line 95% isodose lines.

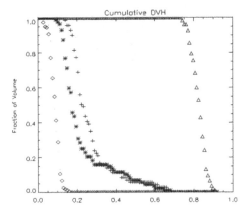

Figure 17.5. Cumulative DVH histogram for the prostate treatment: triangle, prostate; cross, bladder; asterisk, rectum; dot, right femur; diamond, left femur.

The present results represent an improvement with respect to standard radiotherapy treatments. Grant and Woo (1999) have shown that by using standard protocols with sequential tomotherapy, the levels of complication are reduced. Unfortunately, there is not enough information yet to address what the impact of IMRT on the tumor control probability is. With helical tomotherapy, these results should be further improved. The levels of dose obtained in the RAR allow, in principle, new dose escalation protocols or even different dose fractionation schemas.

17.2.2 Megavoltage computed tomography

As was described, the presence of high gradients in IMRT makes necessary the implementation of verification capabilities. For instance in a prostate treatment, the rectum and bladder shapes may change. Therefore, it is highly desirable to have information about the anatomy of these structures on a routine basis. In order to obtain information about organ shape changes, an image modality able to detect low-contrast structures and resolve details of the order of 1 mm is very appealing. Megavoltage computed tomography (MV CT), which uses the treatment linac to generate tomographic images of the patient, should be an adequate tool for this verification. Figure 17.6 is a 3D surface rendering of a rando phantom with MV and kV CT using 12 cGy dose in both cases. The reconstruction method used was filtered back-projection (FBP). Details of the megavoltage CT data preprocessing and reconstruction techniques developed by Ruchala can be found elsewhere (Ruchala 1998, Ruchala *et al* 1999). The quality of the reconstructed images is comparable, and it can be seen that in the MV case the artifacts at the surface due to beam hardening are reduced.

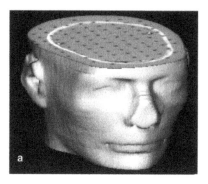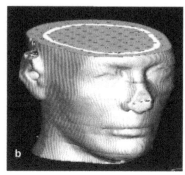

Figure 17.6. (*a*) Three-dimensional surface rendering of a 12 cGy MV CT of the rando phantom. (*b*) Comparison with a 12 cGy kV serial (nonhelical) CT of the rando phantom.

Figure 17.7(*a*) is a high-contrast phantom used to determine the limit of CT resolution. The smallest objects that can be distinguished are of the order of 1 mm. They are air cylinders in a 1 g cm⁻³ density background. Figures 17.7(*b*) and (*c*) show MV CT reconstructions for low-contrast phantoms obtained delivering 175 cGy and 7 cGy respectively. Figure 17.7(*b*) shows that if enough photons are used to obtain the image, the quality is very good and all the contrast regions can be resolved. However, 175 cGy is too high a dose to be used for anatomic verification. Figure 17.7(*c*) is a MV CT with a much smaller dose (2 cGy) deposited that can be considered as adequate for verification. In this case an order of 1.5% contrast can be easily observed.

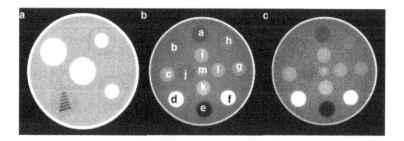

Figure 17.7. (*a*) MV CT of a high-contrast resolution phantom. (*b*) 175 cGy MV CT of a low-contrast phantom. The contrast objects in percent are: a = −3.3, b = −0.4, c = 4.0, d = 17.9, e = −10.0, f = 12.0, g = 2.6, h = 1.0, i = 4.0, j = −1.4, k = 4.8, l = 3.1, m = spectral artifact. (*c*) 7 cGy MV CT of a low-contrast phantom. The contrasts are the same as in (*b*).

From the presented results it can be concluded that MV CT can be used as a verification tool for IMRT and in particular for tomotherapy. Moreover, these images are necessary for dose reconstruction, and the fact that the CT numbers in

MV CT are obtained in the same beam as the treatment is delivered, makes them convenient for computing the dose. Nonetheless, MV CT is not intended as a diagnostic imaging tool.

17.2.3 Registration in projection space

Methodologies for positioning the patient are necessary to guarantee that the treatment is delivered as planned. Standard radiotherapy localizes the patient by using external fiducial marks, which are aligned with laser lights. Another device used to verify patient positioning is the portal image device, which produces 2D projections of the patient similar to a radiograph.

Due to the high gradients present in tomotherapy the accuracy desired is ±2 mm for the translational movements and ±2 degrees for the rotational movements that sometimes are hardly obtained with the methods used in standard radiotherapy.

By using the CT capabilities of helical tomotherapy, it is possible to develop a technique to obtain the six-parameter (three translational, three rotational) patient offset from the registration sinogram (Aldridge 1998, Fitchard *et al* 1998, 1999a, b). The advantage of the registration in projection space versus registration using CT images is that the time-consuming reconstruction processes can be avoided. Moreover, the registration is performed between 2D sinograms instead of 3D image volumes.

Figure 17.8 presents the reference sinogram and the fraction sinogram corresponding to the phantom shown in figure 17.6. Figure 17.8(*a*) is the reference sinogram considered and is at the $x = 0$, $y = 0$ and $z = 0$ cm phantom position. Figure 17.8(*b*) is the fraction sinogram that is obtained by moving the phantom to $x = 0.450$ cm, $y = -0.230$ cm and $z = 0.20$ cm and adding a roll of 2 degrees. The results obtained by the registration are shown in table 17.1. The results are within the tolerance desired in tomotherapy. The technique assumes that the patient is a rigid body. Under this condition the results are very reliable. Studies are under development to address the robustness of the method under organ motion and shape change.

Table 17.1. Results for the registration in projection space.

	x-offset (cm)	*y*-offset (cm)	$r = \sqrt{x^2 + y^2}$ (cm)	*z*-offset (cm)	roll-offset (deg)
Set experimental	0.450	−0.230	0.505	0.20	2.00
Registration result	0.467	−0.225	0.518	0.29	1.83

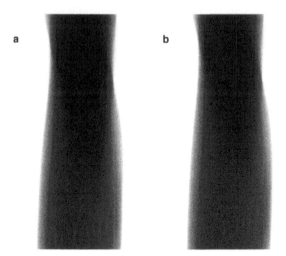

Figure 17.8. (*a*) Reference registration sinogram. (*b*) Fraction registration sinogram.

17.2.4 Delivery modification

After the patient lies on the couch and registration in projection space provides the offset of the patient, there are two possibilities: (1) relocate the patient to the correct position or (2) modify the delivery to the new patient position.

In delivery modification, the plan energy fluence sinogram is changed to modify the delivery according to a rigid (only translations and rotations not, for example, shape changes) patient movement. For instance, if the patient has an offset in the z-direction (i.e. parallel to the axis of rotation), the couch can be translated in that direction to account for that shift. As another example, if the patient is rotated around the axis of the machine, the linac must start the irradiation at a different angular position. If the patient has a more complicated offset, a more complicated transformation can be applied to account for this movement. Therefore, delivery modification is the set of transformations that must be applied to the energy fluence delivery sinogram to account for patient offset. Note that delivery modification is a geometrical transformation of the planned energy fluence pattern that is not very time consuming. A re-optimization will take much longer (Olivera *et al* 2000). Moreover, if the hardware is fast enough to detect patient movement during the fraction, in principle, these changes can be accounted for with delivery modification.

Figures 17.9(*a*) and (*b*) are the optimized and delivery-modified dose distributions for the prostate treatment. The patient's displacement was 1 cm in the direction indicated by the arrow. Both distributions are very similar except for some small differences at the 50% isodose line. Figures 17.9(*c*) and (*d*) are the same dose distributions but overlapped with the regions of interest.

Figure 17.10(*a*) is the energy fluence delivery sinogram for the prostate delivery at the position ($x = 0$, $y = 0$, $z = 0$) cm, and figure 17.10(*b*) is the delivery-modified energy fluence sinogram that has the transformations accounting for a shift of all the voxels along the vector ($x = 1$, $y = 0$, $z = 0$) cm.

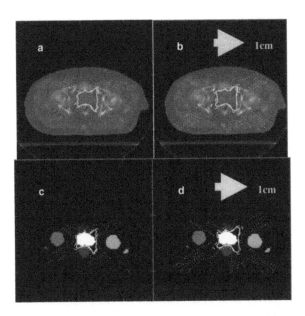

Figure 17.9. (*a*) and (*b*) CT for the optimized and delivery-modified prostate treatment with isodose lines, respectively: dotted line 50%, dashed line 80%, full line 95% isodose lines; (*c*) and (*d*) ROIs for the optimized and delivery-modified prostate treatment with isodose lines, respectively: dotted line 50%, dashed line 80%, full line 95% isodose lines.

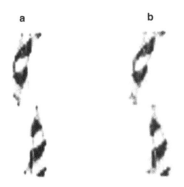

Figure 17.10. (*a*) Planned energy fluence sinogram. (*b*) Delivery-modified sinogram.

As can be observed, delivery modification is a valuable tool to account for patient offset previous to the treatment and may be a very useful tool to account for movements during treatment.

17.2.5 Delivery verification

Once the patient is in the correct position or the delivery has accounted for patient offset, the treatment can be delivered. At this stage the exit detector is collecting the radiation that goes through the patient. This radiation can be used to verify the energy fluence actually delivered to the patient. The dose at the detector i for the projection v can be written as

$$d_i^v = \sum_j D_{ij} w_j \tag{17.2}$$

where D_{ij} is the dose at the detector i per unit of energy fluence at the leaf j (transfer matrix) and w_j is the energy fluence at the leaf j. The generalization of the previous equations for all the detectors in that projection is

$$d^v = D\,w \tag{17.3}$$

where d^v is the vector dose at the detector for the projection v, w is the energy fluence at the MLC for projection v and D is the transfer matrix for that projection. The above equations are valid if the system is linear. Kapatoes *et al* (1999) have shown that there are nonlinearities, such as tongue and groove, leakage and transmission, etc. Those authors provide corrections to make the system linear. Under these conditions, the energy fluence at the projection v can be computed by:

$$w = D^{-1}d^v. \tag{17.4}$$

In summary, delivery verification computes the actual energy fluence delivered to the patient (fluence verification sinogram) from the signal collected at the detector during the delivery (treatment detector sinogram). The energy fluence verification sinogram can be compared with the planned energy fluence sinogram to quantify the quality of the delivery. If during the patient irradiation errors out of tolerance are found, the machine can be shut down.

Figures 17.11(*a*)–(*c*) are the planned delivery, the verification sinogram and the difference between them. As can be observed the delivery is well within tolerance because all the errors are contained in a ±2% interval. In this case, the differences are due to pulse-by-pulse linac fluctuations.

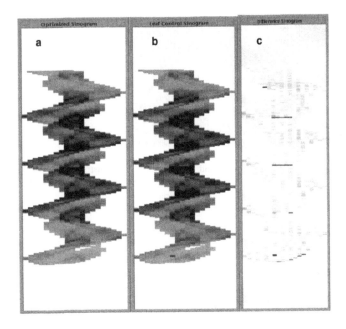

Figure 17.11. (*a*) Planned energy fluence sinogram. (*b*) Verification sinogram. (*c*) Percentage difference between planned and verification sinograms.

17.2.6 Dose reconstruction

Dose reconstruction can be performed if the anatomy of the patient is known during the treatment fraction (McNutt *et al* 1997). A CT just previous to, during or after the delivery provides enough information to verify the patient's anatomy. Using the verification sinogram and the CT done at the treatment fraction the dose actually deposited to the patient can be calculated. This dose reconstruction can be superimposed with the CT to show to the physician an apples-to-apples comparison between planned and actual delivered dose.

Figures 17.12(*a*) and (*b*) are the planned and reconstructed dose distributions computed using the sinograms shown in figure 17.11 for an irradiation of a homogeneous phantom. In this case a U-shaped tumor is irradiated to a high dose and the region inside the U is a sensitive structure. Both dose distributions are very similar, indicating that that the delivery was successful. Figure 17.12(*c*) is a gray scale plot of the difference between planned and reconstructed dose distributions. The differences range between ±1.5%, therefore, the delivery can be considered as clinically acceptable.

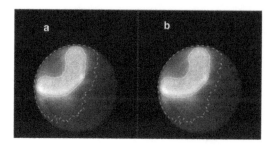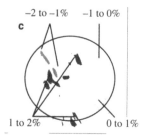

Figure 17.12. (*a*) and (*b*) Planned and reconstructed dose distribution with isodoses lines, respectively: dashed line 30% isodose; dotted line 80% isodose; full line 95% isodose. (*c*) Percentage difference between planned and reconstructed dose distributions.

17.3 CONCLUSIONS

Tomotherapy is a unique therapeutic modality that includes intensity-modulated delivery in conjunction with tomographic verification capabilities. The group of tomotherapy processes just described provides a closed loop for planning, delivery and verification in radiotherapy. The hardware and processes present in a single piece of equipment allows tomotherapy the possibility of playing a major roll in the radiotherapy of the next millennium.

Acknowledgments

We would like to acknowledge the collaboration of the people in the tomotherapy research group and Tomotherapy Inc. Without their participation this project would not have been possible.

PROBLEMS

17.1 What is intensity modulated radiation therapy (IMRT)? What modern devices can be used to perform IMRT?

17.2 What are the main pieces of hardware in a tomotherapy machine?

17.3 What are the processes used for tomotherapy?

17.4 Why is it important that the set of tomotherapy processes forms a closed loop for planning, delivery and verification of radiotherapy?

17.5 Why is improved accuracy in positioning and delivery necessary for tomotherapy?

17.6 Why is an objective function used for treatment planning optimization? What are the two main types of objective function that can be used in treatment planning optimization?

17.7 Why is megavoltage computed tomography (MV CT) an important process for tomotherapy verification?

17.8 What are the advantages of *patient registration* in projection space as opposed to megavoltage CT for patient positioning?

17.9 Why might the process of *delivery modification* be preferred over physically repositioning the patient?

17.10 What is *delivery verification*?

17.11 Why is *dose reconstruction* a key processes in the evaluation of a patient treatment?

REFERENCES

Aldridge J S 1998 Patient registration and conformal avoidance radiotherapy *Prospectus PhD Thesis* Dept Medical Physics, University of Wisconsin–Madison

Balog J 1996 Helical delivery of intensity modulated external beam radiotherapy *PhD Thesis* Dept Medical Physics, University of Wisconsin–Madison

Balog J 1998 Tomotherapy dosimetry and the tomotherapy benchtop *PhD Thesis* Dept Medical Physics, University of Wisconsin–Madison

Balog J P, Mackie T R, Reckwerdt P, Glass M and Angelos L 1999a Characterization of the output for helical delivery of intensity modulated slit beams *Med. Phys.* **26** 55–64

Balog J P, Mackie T R, Wenman D L, Glass M, Fang G and Pearson D 1999b Multileaf collimator transmission *Med. Phys.* **26** 176–86

Bortfeld T, Boyer A L, Schlegel W, Kahler D L and Waldron T J 1994 Realization and verification of three-dimensional conformal radiotherapy with modulated fields *Int. J. Radiat. Oncol. Biol. Phys.* **30** 899–908

Brahme A 1988 Optimization of stationary and moving beam radiation therapy techniques *Radiother. Oncol.* **12** 129–40

Carol M, Grant W H, Bleier A R, Kania A A, Targovnik H, Butler E B and Woo S W 1996 The field-matching problem as it applies to the peacock three-dimensional conformal system for intensity modulation *Int. J. Radiat. Oncol. Biol. Phys.* **34** 183–7

Carol M, Targovnik H, Smith D and Cahill D 1992 3D planning and delivery system for optimzed confomal therapy *Int. J. Radiat. Oncol. Biol. Phys.* **24** 159

Cormack A M 1987 A problem in rotation therapy with X-rays *Int. J. Radiat. Oncol. Biol. Phys.* **13** 623–30

Cormack A M and Cormack R A 1987 A problem in rotation therapy with X-rays 2: dose distributions with an axis of symmetry *Int. J. Radiat. Oncol. Biol. Phys.* **13** 1921–5

Fitchard E E, Aldridge J S, Reckwerdt P J, Iosevich A, Olivera G H and Mackie T R 1999a Six parameter patient registration directly from projection data *Nucl. Instrum. Methods A* **421** 342–51

Fitchard E E, Aldridge J S, Reckwerdt P J and Mackie T R 1998 Registration of synthetic tomographic projection data sets using cross-correlation *Phys. Med. Biol.* **43** 1645–57

Fitchard E E, Aldridge J S, Ruchala K, Fang G, Balog J P, Pearson D W, Olivera G H, Schloesser E A, Wenman D, Reckwerdt P J and Mackie T R 1999b Registration using tomographic projection files *Phys. Med. Biol.* **44** 495–507

Grant W III and Woo S Y 1999 Clinical and financial issues for intensity modulated radiation therapy delivery *Semin. Radiat. Oncol.* **9** 1–10

Gustafsson A, Lind B K and Brahme A 1994 A generalized pencil beam algorithm for optimization of radiation therapy *Med. Phys.* **21** 343–56

Kapatoes J M, Olivera G H, Reckwerdt P J, Fitchard E E, Schloesser E A and Mackie T R 1999 Delivery verification in sequential and helical tomotherapy *Phys. Med. Biol.* **44** 1815–41

Ling C C, Burman C, Chui C S, Kutcher G J, Leibel S A, LoSasso T, Mohan R, Bortfeld T, Reinstein L, Spirou S, Wang X H, Wu Q, Zelefsky M and Fuks Z 1996 Conformal radiation treatment of prostate cancer using inversely-planned intensity-modulated photon beams produced with dynamic multileaf collimation *Int. J. Radiat. Oncol. Biol. Phys.* **35** 721–30

Low D A 1997 Abutment region dosimetry for sequential arc IMRT delivery *Phys. Med. Biol.* **42** 1465–70

Low D A and Mutic S 1998 A comercial IMRT treatment planning dose calculation algorithm *Int. J. Radiat. Oncol. Biol. Phys.* **41** 933–7

Mackie T R 1997 Tomotherapy *XII Int. Conf. Use Computers Radiat. Therapy (Salt Lake City, UT)* ed D D Leavitt and G Starkshall (Madison, WI: Medical Physics) pp 9–11

Mackie T R *et al* 1997 Tomotherapy, rethinking the process of radiotherapy *XII Int. Conf. Use Computers Radiat. Therapy (Salt Lake City, UT)* ed D D Leavitt and G Starkshall (Madison, WI: Medical Physics)

Mackie T R, Balog J, Ruchala K, Shepard D, Aldridge J S, Fitchard E E, Reckwerdt P, Olivera G H, McNutt T and Mehta M 1999 Tomotherapy *Semin. Radiat. Oncol.* **9** 108–17

Mackie T R, Holmes T, Swerdloff S, Reckwerdt P, Deasy J O, Yang J, Paliwal B and Kinsella T 1993 Tomotherapy: a new concept for the delivery of dynamic conformal radiotherapy *Med. Phys.* **20** 1709–19

McNutt T R, Mackie T R and Paliwal B R 1997 Analysis and convergence of the iterative convolution/superposition dose reconstruction technique for multiple beams and tomotherapy *Med. Phys.* **24** 1465–76

Mohan R, Lovelock M, Mageras G, LoSasso T and Chui C 1996 Computer-controlled radiation therapy and multileaf collimation *Frontiers Radiat. Ther. Oncol.* **29** 123-38

Oldham M, Khoo V S, Rowbottom C G, Bedford J L and Webb S 1998 A case study comparing the relative benefit of optimizing beam weights, wedge angles, beam orientations and tomotherapy in stereotactic radiotherapy of the brain *Phys. Med. Biol.* **43** 2123–46

Oldham M and Webb S 1997 Intensity-modulated radiotherapy by means of static tomotherapy—a planning and verification study *Med. Phys.* **24** 827–36

Olivera G H, Shepard D M, Ruchala K, Aldridge J S, Kapatoes J M, Fitchard E E, Reckwerdt P J, Fang G, Balog J, Zachman J and Mackie T R 1999 *Tomotherapy Modern Technology of Radiation Oncology* ed J Van Dyk (Madison, WI: Medical Physics) pp 521–87

Olivera G H, Fitchard E E, Reckwerdt P J, Ruchala K and Mackie T R 2000 Delivery modification as an alternative to patient repositioning in tomotherapy *XIII Int. Conf. Use Computers Radiat. Therapy (Heidelberg, Germany)* ed W Schlegel and T Bortfeld (New York: Springer) pp 294–6

Purdy J L 1996 Intensity-modulated radiation therapy *Int. J. Radiat. Oncol. Biol. Phys.* **35** 845–6

Ruchala K 1998 Megavoltage computed tomography for tomotherapy verification *Prospectus PhD Thesis* Dept Medical Physics, University of Wisconsin–Madison

Ruchala K J, Olivera G H, Schloesser E A and Mackie T R 1999 Megavoltage CT on a tomotherapy system *Phys. Med. Biol.* **44** 2597–621

Shepard D M, Olivera G H, Angelos L, Sauer O, Reckwerdt P and Mackie T R 1999 A simple model for examining issues in radiotherapy optimization *Med. Phys.* **26** 1212–21

Shepard D M, Olivera G H, Reckwerdt P and Mackie T R 2000 Iterative approaches to dose optimization in tomotherapy *Phys. Med. Biol.* **45** 69–90.

Spirou S V and Chui C S 1994 Generation of arbitrary intensity profiles by dynamic jaws or multileaf collimators *Med. Phys.* **21** 1031–41

Sternick E S 1997 *The Theory and Practice of Intensity Modulated Radiation Therapy* (Madison, WI: Advanced Medical)

Svensson R, Kallman P and Brahme A 1994 An analytical solution for the dynamic control of multileaf collimators *Phys. Med. Biol.* **39** 37–61

Swerdloff S, Mackie T R and Holmes T 1994 *Multi-leaf radiation attenuator for radiation therapy* US patent 5,351,280 van Santvoort J P and Heijmen B J 1996 Dynamic multileaf collimation without 'tongue-and-groove' underdosage effects *Phys. Med. Biol.* **41** 2091–105

Verhey L J 1996 3D conformal therapy using beam intensity modulation *Frontiers Radiat. Ther. Oncol.* **29** 139–55

Webb S 1991 Optimization by simulated annealing of three-dimensional conformal treatment planning for radiation fields defined by multileaf collimator *Phys. Med. Biol.* **36** 1201–26

Webb S 1997 *The Physics of Conformal Radiotherapy: Advances in Technology* (Philadelphia, PA: Institute of Physics)

Xia P and Verhey L J 1998 Multileaf collimator leaf sequencing algorithm for intensity modulated beams with multiple static segments *Med. Phys.* **25** 1424–34

Yu C X, Symons M J, Du M N, Martinez A A and Wong J W 1995 A method for implementing dynamic photon beam intensity modulation using independent jaws and a multileaf collimator *Phys. Med. Biol.* **40** 769–87

CHAPTER 18

DRUG DELIVERY

John R Goomey

The most common method of medical therapy is the administration of drugs. Billions of dollars are spent annually on the development of safer and more effective medicines. Considerable efforts are also made in enhancing the methods used to deliver these drugs. Commonly medicine is delivered to the body through pills, inhalers and injections. Goals in finding new delivery methods include: reducing systemic effects on the body, increasing localized effects, reducing the need for needle injections and minimizing the burden for those who need regularly administered drugs, such as diabetics.

18.1 NONINVASIVE DRUG DELIVERY

Noninvasive drug delivery, for the purpose of this section, is the administration of medicine without breaking the skin on a macroscopic level. Most medicine is delivered in this way orally, through pills and inhalers, however several methods are being developed so that many injected medicines may also be delivered without the pain and risks involved in using traditional needles. These include the employment of electric fields, chemicals, ultrasound, pressure and microfabricated needles to drive medicine through the skin.

18.1.1 Respiratory delivery

Most people have taken medicine orally, in the form of pills, liquids or vapors. Though these methods are effective, improvements in efficiency of delivery are desirable. Metered dose inhalers, for example, provide an atomized vapor to be inhaled. The effectiveness of the delivery is hindered by several problems stemming from the short burst of vapor that is administered. Due to the short duration of delivery, the timing with which the patient inhales greatly affects the amount of medicine delivered to the lungs, along with the distribution of medicine on the lining of the lung. In addition, the force with which the medicine comes out often results in much of the drug being deposited in the mouth and trachea before reaching the lungs. In spite of this, administering medicine through the employment of inhalers is an attractive method of delivery. It is simple, painless, for respiratory therapy provides direct delivery to the system in need of medicine, and for other therapies provides a quick pathway to the blood without injection.

To deliver medicine effectively through inhalation, the inhalation must be slow and deep. In addition, the medicine must be in small particles for effective delivery, from 1 to 3 μm in diameter. If the particles are too large, they are absorbed in the upper respiratory tract, and if they are too small, they are exhaled. Particles larger than 2.5 μm cannot pass the air–blood barrier of the lung.

The common technologies for inhaled delivery of medicine are metered dose inhalers, dry powder inhalers and nebulizers. Metered dose inhalers are pressurized containers of medicine, which deliver an aerosol containing the medicine for delivery. They are very common, due to their convenience of use, low cost and portability. The problems with metered dose inhalers are outlined above. Dry powder inhalers are breathing-activated inhalers that deliver micronized powders of medicine. These overcome the problems associated with the sudden burst delivered by metered dose inhalers, but tend to provide a less effective dose of medicine due to clumping of the particles. Nebulizers provide a constant atomized mist of medicine. A gentle *steam* is provided, which is inhaled through normal breathing. Nebulizers solve the major problems of inhaled delivery as far as effectiveness, however they tend to be bulky, expensive and require an electric outlet for continued operation. Significant efforts have been and are being made to overcome the problems with each of these delivery systems.

18.1.1.1 Metered dose inhalers. There have been only two real advances in metered dose inhalers: the introduction of nonchlorofluorocarbon propellants, and the introduction of a spacer device to buffer the burst of aerosol. The spacers are quite effective in increasing the effective dose of medicine, though they add some bulkiness to the inhaler, reducing the convenience and portability.

18.1.1.2 Dry powder inhalers. Many systems have been and are being developed to overcome the problems with dry powder inhalers. The main disadvantages of dry powder inhalers result from the clumping of the particles, and the need to load each dose into the inhaler (as opposed to the metered dose inhalers, which contain multiple doses). Dry powder inhalers are propellant free and only deliver medication when the patient inhales sharply to disperse the particles for distribution in the lungs. Even with sharp inhalation, the amount of force required to break apart the clumps of particles is not necessarily supplied, resulting in poor delivery efficiency. Several companies have devised methods for improving dry powder inhalers. One such company is Inhale Therapeutic Systems, which has developed a device to eliminate the problem of clumping. The device uses a burst of compressed air to aerosolize the powder and produce a standing cloud in a holding chamber, from which the patient can then inhale. The aerosolized powder, along with a backing pressure from the burst of air, allows for effective delivery of powdered medicines to the lungs.

18.1.1.3 Nebulizers. Nebulizers provide the most effective means of delivering medicine through the lungs. They solve the delivery problems of the other methods, but suffer from other disadvantages. Nebulizers use buffered aqueous solutions of medicine. Since they use aqueous solutions, they must be kept clean and sanitized. Failure to do this can result in the delivery of bacteria deep into the lungs. Practically they are bulky, inconvenient and expensive.

There are two primary types of nebulizer: air-jet and ultrasonic. Figure 18.1 shows that the air-jet nebulizers force a steam of compressed air over the surface of the aqueous solution. A low-pressure region, caused by the Venturi effect, draws the solution up a capillary tube. The pressure differential and wind shear produces small droplets of solution. Baffles are used to condense drops that are too large, and the vapor escapes so that it is available for inhalation.

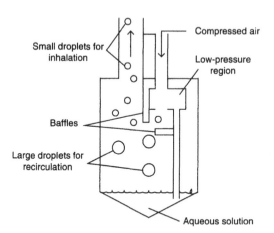

Figure 18.1. In an air-jet nebulizer the solution is drawn up a capillary tube by a pressure differential, and broken up into small droplets by the shear forces of the airflow. Small droplets escape, and large droplets are condensed and re-circulated.

Ultrasonic nebulizers use piezoelectric crystals to direct high-frequency waves through the aqueous solution of medicine, producing a vapor for inhalation. The size of piezoelectric-driven nebulizers allows for smaller, more portable nebulizers. The piezoelectric models are not yet as robust as the air-jet models. In addition, they are more expensive.

18.1.2 Transdermal delivery

The delivery of medication through the skin, without the need for a needle, is an attractive and practical method of drug delivery. The obvious attraction is avoiding the pain involved in receiving shots. There are several advantages associated with transdermal delivery, including the ability to administer a drug slowly and constantly over a long period of time, the ability to bypass the digestive tract, which can deactivate or reduce the effectiveness of some drugs, a minimized risk from blood contamination from used needles and a greatly enhanced ability to stop the delivery of the drug.

18.1.2.1 The skin barrier. The skin provides, for the large part, an impervious barrier. Though some compounds readily absorb through the skin, most do not. In order to overcome this barrier several methods have been developed. These include the employment of electric fields, chemicals, ultrasound, pressure and microfabricated needles to drive medicine through the skin.

Figure 18.2 shows that the skin is comprised of two primary layers, the epidermis and the dermis. The dermis is in close contact with blood vessels, and does not provide a barrier to the diffusion of drugs through the skin. The epidermis consists of two regions, the viable epidermis and the stratum corneum. The viable epidermis is composed of cells that change in function throughout, with part of its total function being to generate the stratum corneum, the outermost layer of skin.

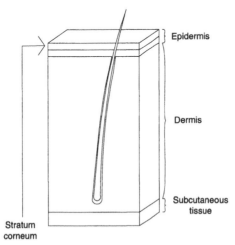

Figure 18.2. The skin consists of two layers, the thick dermis and the epidermis. The epidermis is further divided into two regions, the viable epidermis and the stratum corneum.

The major resistance to drug penetration through the skin is provided by the stratum corneum. Figure 18.3 shows that the stratum corneum, approximately 15–20 μm thick, consists of interlocking flattened dead skin cells (corneocytes), which consist primarily of keratin. The area between the corneocytes is filled with an assortment of lipids.

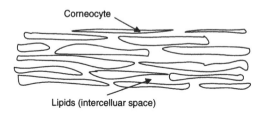

Figure 18.3. The interlocking arrangement of corneocytes in the stratum corneum.

Hair follicles and sweat glands provide one bypass to the stratum corneum. Though it has been claimed that these pathways provide the easiest route through the skin, studies indicate that the intercellular lipids provide the pathway for drugs to penetrate the skin.

Transport through the skin is a diffusion process, and is described by Fick's first law of diffusion. The steady-state flux of the drug through the skin J is given by

$$J = -D \cdot A \frac{\partial c}{\partial x} \tag{18.1}$$

where D is the diffusion coefficient of the drug (essentially a conductivity), A is the area and $\partial c/\partial x$ is the concentration gradient (or driving force). The diffusion coefficient of the drug is, for the most part, the controlling factor in transport through the skin. The area cannot be increased without limit, and the concentration gradient cannot be increased significantly due to the risk of overdosing. The driving force, in the simple case, is the concentration gradient. Transdermal delivery systems attempt either to enhance the driving force, through means of pressure or electric fields, or to enhance the diffusion through use of ultrasound or chemical means. A couple of systems attempt to breech the stratum corneum painlessly in order to deliver the medication.

18.1.2.2 Transdermal patch. The most familiar and readily available transdermal system is the transdermal patch. This system consists of an adhesive patch that delivers a controlled dose of a drug through the skin over a long period, usually up to a week. The most familiar transdermal patch is the nicotine patch, used to aid in smoking cessation programs. Several therapies are available employing transdermal patches, treating such conditions as nicotine addiction, travel sickness, arthritis, pain relief, angina and hormone replacement therapies.

Figure 18.4 shows that the basic components of a transdermal patch are a drug reservoir, a permeable membrane and an adhesive to attach the patch to the skin. An alternative system consists of only two layers, a backing and an adhesive. In these systems the adhesive, drug and penetration enhancer are combined into a matrix. Patches are placed on the body in places where the stratum corneum is thinnest (usually corresponding to where the skin is softest). Commonly the inner forearm, upper arm or back is selected.

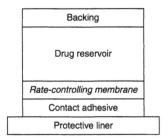

Figure 18.4. The basic transdermal patch consists of a contained drug reservoir and a contact adhesive. Some systems employ a rate-controlling membrane to control the rate of delivery to the skin, while others use the skin as the rate-controlling membrane.

The primary design considerations in transdermal patches are the adhesive properties and transport enhancement through the skin. Though adhesion to the skin may seem trivial, as adhesive bandages are common, additional considerations must be taken into account with transdermal drug delivery systems. The adhesive cannot irritate the skin, it must be able to stick for up to seven days, it must be easily removable, it cannot make the skin less permeable, it must not interact with the drug being delivered, the drug being delivered must be soluble in the adhesive and its properties cannot change drastically when saturated with the drug. A detailed discussion of the requirements for transdermal systems can be found in a review by Venkatraman and Gale (1998).

18.1.2.3 Transport enhancement. The key to all transdermal delivery is the enhancement of transport through the skin. In the case of transdermal patches, the method that enhances the transport is, for the most part, limited to a chemical method. Penetration enhancers are often included in the drug solution to aid or enable transport. Since the pathway for the drugs is the intercellular lipids, it is necessary to break down the rigid barrier the lipids provide to transport most drugs through the skin. The chemicals used to enhance transport through the lipids can be classified in two main groups: solvents and surfactants. Solvents act to break down the structure of the lipids, creating a more jelly-like structure. Surfactants are chemicals that temporarily diminish the barrier resistance.

Solvents, in order to break down the intercellular lipids, must have a similar structure to the lipids. Though simple alcohols and common organic solvents can be effective in interrupting the lipid barrier, possible toxic effects from the more common compounds make them undesirable for use in drug delivery systems, though some are used in commercial systems. Examples of such compounds are methanol, propylene glycol and acetone. Solvents used as penetration enhancers include ethanol, oleic acid, decyl methyl sulphoxide, propylene glycol or laurocapram (Azone®). These compounds impart structural disorder to the lipids. Decyl methyl sulphoxide is only truly effective in high concentrations. In addition, the disruption of the stratum corneum is permanent and only reversed when the skin redevelops. Laurocapram was the first chemical specifically designed for enhancing transport through the skin.

Surfactants are chemicals that affect the surface tension of a substance. In the case of transdermal enhancement, surfactants perform the function of disturbing the lipid cell membranes. The result is an increased fluidity in the lipids allowing for transport enhancement. Surfactants are used sparingly as enhancers in patches, but are generally used to enhance iontophoresis, or electric transport through the skin.

Other chemicals have been shown to enhance transport through the skin. Urea, calcium thioglycolate and eucalyptol are among them. Several other compounds have the ability to enhance transport, but the mechanisms and usefulness of the compounds are still being researched.

18.1.2.4 Safety and limitations. The benefits of some transport enhancers must be balanced by potential problems and health threats. Many of the transport enhancers can be harmful to the body in large concentrations, as may occur in the case of long-term treatment. For example, though urea has the ability to enhance the absorption of hydrocortisone through the skin, there is question as to the potential toxicity of long-term exposure, as is the case with some of the enhancers commonly used. Manufactured enhancers, therefore, are designed with additional criteria taken into account. It is desirable for the enhancers to remain in the stratum corneum, rather than enter the bloodstream. In order to accomplish this, the enhancers are designed to be lipophilic (attracted to lipids), and hydrophobic (repulsed by water). This tends to keep the enhancers in the outermost layer of skin, and out of the general system.

Other issues with transport enhancers are the selectivity of drugs the enhancer is useful for, and the hindrances to drug transport some of them cause. In the first case, the method of transport enhancement from chemical to chemical varies slightly, making each enhancer useful for a limited class of drugs. In the second case, some transport enhancers decrease the effective concentration gradient of the drug being delivered, thus reducing the driving force. Though the ability of the drug to diffuse through the skin may be increased, the reduction in the driving force often cancels the benefits of the increased diffusivity.

An alternative to transport enhancers is suggested in a review by Hadgraft and Pugh (1998). Most drugs have been designed with the intention of conventional administration. For example, many drugs are optimized for absorption in the gastrointestinal tract. These drugs may not be in the best form for being delivered transdermally. A better approach may be to design drugs that are specifically meant to be delivered transdermally and act as their own enhancers. This would eliminate many of the problems with enhancers, and possibly broaden the capability of patches.

As with any patient-administered system, patient compliance is an important issue to usefulness. Several patch systems on the market today cause skin irritation in some patients. Many of the transport-enhancing chemicals cause significant skin irritation with long-term use. Since the patches cause discomfort, many patients discontinue using them. Discontinued use of a medication can often be more harmful, so for patch systems to gain widespread use for a significant number of medications, the discomfort caused by the patches must be reduced.

18.1.2.5 Microfabricated needles. Recent development in silicon processing due to advancement in the field of microelectromechanical systems (MEMS) has led to new possibilities for transdermal delivery of drugs. Figure 18.5 shows that using a reactive ion etch process, Henry *et al* (1998) created an array of 400 needles in a 3 mm × 3 mm region. Though the needles are solid, and therefore incapable of actually injecting drugs, they are capable of breaking through the stratum corneum mechanically, allowing drug delivery through the breach. The group claims that limited human tests demonstrated that the needle array was painless, and volunteers described the sensation as being similar to the application of adhesive tape.

Figure 18.5. An array of microneedles, each approximately 150 μm long (courtesy of Georgia Institute of Technology).

A fluorescent molecule, calcein, was used to model the permeability enhancement. Typical diffusion rates of calcein through untreated skin are less than 1×10^{-12} cm h^{-1}. With the needles inserted, the diffusion increased by three orders of magnitude. Inserting and then removing the needle array increased permeability by more than four orders of magnitude. Ongoing research may result in an array of needles such as these that are hollowed out, allowing for painless injection of large molecules through the skin. Even without this advance, however, a three-to-four order of magnitude increase in transdermal transport is an impressive and practical increase.

18.1.2.6 Iontophoresis and electroporation. Physical means as well as chemical means may enhance transdermal transport. Iontophoresis is a method in which an ionized drug is driven through the skin by means of a small electric current. Iontophoresis has been shown as an effective means to transport charged species, as well as water-soluble neutral species through the skin. Many of the compounds that are difficult to transport through the skin via chemical enhancement transport much more efficiently using iontophoresis. This is primarily because the more difficult species to transport through the skin by means of conventional enhancement are compounds that are charged or polar.

Electroporation differs from iontophoresis in that instead of a low-level, steady direct current being applied, high-voltage pulses are used. Though the set-up of the system is analogous, both the applied signal and mode of transport differ from iontophoresis.

Figure 18.6 shows that iontophoretic transport of drugs through the skin is accomplished by applying small electric currents (< 5 mA) through an electrode filled with medicine. The polarity of the electric signal is determined by the polarity of the molecule to be delivered through the skin. The molecule to be delivered is stored in the like-charged terminal of the delivery system. The oppositely charged terminal (with opposite charge to the molecule being delivered) is placed at a distance from the drug source. The electric field between the probes causes the molecule to flow through the skin toward the oppositely charged terminal. While in transit through the dermis, the drug is in contact with blood vessels and can be carried away to the surrounding tissue. In order to analyze the iontophoresis, an understanding of the electrical behavior of the skin is necessary.

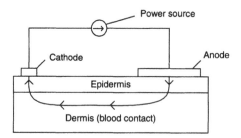

Figure 18.6. Schematic showing the principle of iontophoresis. A pair of electrodes placed on the skin create an electric field in the upper region of the skin. The anode contains a reservoir with an ionized drug that migrates through the epidermis into the dermis where it may enter the bloodstream. In the case of electroporation, the current supply would be replaced with a high-voltage, pulsed supply.

18.1.2.7 Electrical behavior of skin. In the case of iontophoretic flux through the skin, the steady-state flux ($J_{\Delta\psi}$) is given by the modified Nernst–Planck model

$$J_{\Delta\psi} = -\varepsilon\left[D\left(\frac{dC}{dx} + \frac{CzF}{R_gT}\frac{d\psi}{dx}\right) \pm vC\right] \qquad (18.2)$$

where ε is the combined porosity and tortuosity, D is the diffusion coefficient of the permeant, ψ is the electric potential, R_g is the gas constant, C is the concentration of the permeant, x is the position of the membrane, z is the charge number of the permeant, F is the Faraday constant, T is the temperature and v is the average velocity of the convective flow. The first term in the parentheses is just that of nonenhanced transport as in equation (18.1). The second term in the parentheses takes into account the ionic nature of the permeant and the voltage applied. The last term is related to the phenomena of electro-osmosis. Electro-osmosis is the flow of water due to the applied electric field. This equation is a satisfactory first-order explanation of iontophoretic flux, however reality is much more complex. The flux of ions can modify the resistance of skin, which is due to the permeability of mobile ions. This results in a nonlinearity in the equation.

The nonlinearity, in practice, is not a hindrance to accurate monitoring of iontophoretic flux. If we assume that the imposed flux due to the ionic permeant is much greater than the passive diffusion, as is the purpose of iontophoresis, the equation predicts a linear relationship between the flux and the applied voltage. All that is necessary to determine the iontophoretic flux is determining the proportionality constant between the flux and applied voltage. This can be easily accomplished by measuring the impedance of the skin during the administration of the drug.

An important consideration in predicting the effectiveness of iontophoretic delivery is analyzing the impedance of the skin. Early studies of the electrical behavior of the skin revealed that the impedance is complex. Frequency analysis of the impedance of the skin reveals that at very low frequencies, the impedance is nearly constant and it decreases as the frequency of the applied field is increased. This indicates that the skin is capacitive in nature. Attempts to create a simple model for the impedance of the skin, consisting of a resistance in series with a variable complex impedance, were inaccurate. It was found that to accurately model the skin, at least two parallel resistance–capacitance (*RC*) circuits in series were necessary to model the skin. Figure 18.7 shows an accurate model, proposed by R T Tregear in 1966, which consisted of a series of parallel *RC* circuits, with a small series resistance, representing the substratum corneum layers. The analogy represents each of the layers of cells in the stratum corneum as a parallel *RC* circuit. In 1971, this model was tested by D T Lykken, and proven accurate in modeling skin impedance.

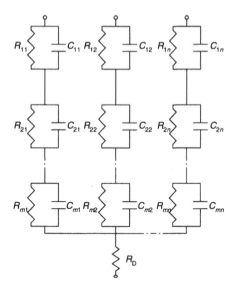

Figure 18.7. Equivalent circuit model for the skin showing the analogy between the parallel resistive–capacitive circuits and the cell layers of the stratum corneum. The resistances and capacitances with numeric subscripts represent the area and layers of the stratum corneum. R_D represents the resistance of the tissue below the stratum corneum.

The impedance of the skin also decreases over time with continued applied electric field. Though this decrease is not completely understood, recent studies seem to imply that new conductive pathways are created through the stratum corneum. From the above discussion it might be inferred that using alternating current (ac) waveforms in iontophoresis would result in greater permeant flux. This is not the case. Although the impedance of the skin is greatly reduced, transport is not increased, partially due to the alternating field not producing a net flux in any direction. In addition, the alternating current does not break down the barrier provided by the stratum corneum efficiently. Direct current (dc) voltage is effective in breaking down the skin barrier, and allowing for transport enhancement. Higher voltages result in higher permeability.

18.1.2.8 Practical electrically assisted transport. The transport mechanisms through the skin differ for iontophoresis and electroporation. Although the exact mechanisms are still being studied, the common assumption seems to be that iontophoretic transport occurs primarily by modifying the existing pores and ducts through the skin, such as hair follicles and sweat glands, while electroporation results from a breakdown of the intercellular lipids (Banga and Prausnitz 1998, Gallo *et al* 1997, Li *et al* 1997, 1998).

The crux of making electrically assisted transport widely accepted is to obtain enhanced transport, while keeping the procedure painless and comfortable. For simple effectiveness, the application of high dc voltages is the most effective means of electrically assisted transport. This, however, is harmful to the skin. Even moderate dc voltages, of the order of a few volts, applied over time cause discomfort to the patient, likely the result of polarization effects in the skin. The use of high voltages, even for short periods, can also result in discomfort and damage to the skin. Long-duration pulses at moderate voltage (50 V for 200 ms) enhance delivery more than high-voltage short pulses (100 V for 1 ms), however heating effects on the skin make this impractical.

The optimal solution seems to lie in using a combination of electroporation and iontophoretic techniques. The two effects on the skin from electrical stimulation that enhance transport are the creation of new pathways through the skin and enlarging the pathways that exist. The former is crucial to get any compound through the skin. The later is necessary to allow for drugs, many of which are large molecules, to get through the skin. The technique that seems to hold the most promise is using a pretreatment of several high-voltage pulses prior to iontophoresis. Banga and Prausnitz (1998) mention a study in which a high-voltage pulse was applied prior to iontophoresis, resulting in a doubling of transport. Li *et al* (1997) also demonstrated that applying a moderate-voltage pulse prior to iontophoresis results in a higher conductance through the skin. In addition, superimposing a small ac signal on a low-voltage dc signal has been shown to increase transdermal flux (Li *et al* 1998).

A correlation between the resistance of the skin and the size of the pathways through the stratum corneum has been demonstrated. This implies that as the electric resistance of the skin is decreased, larger molecules can be passed through the skin using iontophoresis. The passage of large molecules, such as insulin, through the skin greatly increases the attractiveness of iontophoresis for drug delivery. Langkjaer *et al* (1998) using animal skin have shown a 1000-fold increase in transdermal flux of insulin using iontophoresis, coupled with skin pretreatment. Zakzewski *et al* (1998) used pulsed-mode iontophoresis in combination with different skin preparations for insulin delivery to rats. Effectiveness was monitored by blood glucose levels in the test subjects. The study showed that blood glucose levels can be affected at therapeutic levels using iontophoresis, providing appropriate preparation is made to the skin.

Although there are several commercial systems available, practical employment of electrically assisted transport is limited. There are several practical problems with iontophoretic systems for delivery of medicine. Long times are necessary (usually over an hour) for significant concentrations of drug to pass through the skin. For a practical system for humans, the system must use pulsed dc, or very low-voltage dc, to reduce or eliminate discomfort. For many molecules, the enhancement, however large, does not result in a level of delivery to make the system practical. Few systems are portable, which would be necessary for them to gain widespread use, based on the time it takes to deliver the drug. Much work is being done in the area of electrically assisted transport, especially in coupling iontophoresis and electroporation with traditional transport enhancers. The combination of including transport enhancers and miniaturizing the iontophoretic systems may result in their presence being common in drug delivery.

18.1.2.9 Sonophoresis. Using sound waves can assist transdermal transport of medication. This technique is called sonophoresis or phonophoresis. This technique has been in use for more than 40 years as a complement to topical application of some medications, and is commonly used in sports medicine today. Coupling a drug to the skin using sound waves at frequencies ranging from 20 kHz to 3 MHz has been shown to dramatically increase the transport of some drugs through the skin. Typically, a coupling gel is applied to the skin, and drugs are transported through the gel into the skin.

Three basic physical effects of ultrasound can potentially lead to increased transport through the skin. Due to increases in localized pressure, the power input from ultrasound can result in localized increases in temperature. Increased temperature can have remarkable effects on diffusivity of compounds through a material. Acoustic streaming effects could increase the concentration of the drug at the skin surface, and the drug could be physically driven through the skin. Cavitation effects (the creation of small bubbles) caused by ultrasound can disrupt the stratum corneum, allowing drugs to penetrate the skin more readily.

Some studies have demonstrated that the transport of compounds through the skin is enhanced by using higher frequencies and higher power densities. This would imply that localized heating and streaming effects are the main mechanisms of transport increase. Although this increase is seen, the magnitude of the enhancement is only of the order of a factor of 10, and suggests that this is not the main mode of transport enhancement. The use of 20 kHz ultrasound has resulted in an increased transport by a factor of more than 1000 (Mitragori *et al* 1996). The dominant physical effect at low frequencies is cavitation; therefore, it can be inferred that the dominant mode of transport enhancement is the temporary breakdown of the intercellular lipids. The mechanical defects caused by kilohertz range ultrasound have been seen in recent studies. Defects in the size range of 20 µm were observed in human skin after the application of ultrasound (Wu *et al* 1998). In this size range, the defects may allow for the transport of large molecules through the skin.

Since the dominant enhancement mode of ultrasound is by cavitation of the stratum corneum, transport is greatly affected by the properties of the drug being applied. Mitragori *et al* (1997) quantified the differences in transdermal transport. Based on the passive permeability, P_p, and octanol–water partition coefficient of a drug, $K_{o/w}$, the estimated sonophoretic transport enhancement e is given by

$$e \sim \frac{K_{o/w}^{0.75}}{(4 \times 10^4) P_p}. \tag{18.3}$$

Although the equation is approximate, the relationship does agree well with experimental values and observations. For example, the transport enhancement of testosterone ($P_p = 2.2 \times 10^{-3}$ cm h^{-1}, $K_{o/w} = 2070$) was measured to be $e = 4 \pm 1.1$, and the calculation gives $e = 3.5$.

Ultrasonic transport enhancement seems promising as a technique for transdermal drug delivery. However, the mechanisms of transport enhancement and the usefulness of the approach are still under debate. Often less expensive equipment, such as iontophoretic set-ups, can produce similar results, making it a more attractive technique for researchers to study. Advantages of ultrasound delivery can be found for some drugs, especially in dermatology, and the safety of ultrasound at low powers has been thoroughly demonstrated. The area in which ultrasound may have the strongest capability to be useful is in the area of controlled-release delivery.

18.1.2.10 Jet injection. A slightly more forceful technique for transdermal delivery is a jet injector system. This method is used in conjunction with a transdermal delivery system, such as a patch. Prior to application of the medicine, a jet injector filled with saline is applied to the skin. This creates a pore in the stratum corneum through which drugs can travel into the dermis (Inoue *et al* 1996). This technique breeches the skin in a similar way to a needle, without actually penetrating the skin with an instrument. Though the technique provides greatly enhanced transdermal transport for up to a week, the initial treatment is similar to a shot, thus is not likely to gain widespread acceptance by patients.

18.1.3 Oral controlled-release delivery

One aspect of controlled delivery is that of sustained constant dose. This can be achieved through the employment of transdermal patches, implants or the slow release of a drug into the system by other means. The noninvasive approaches are the transdermal patches and time-released drugs. Implants are discussed in the following sections. The need for time-controlled release of drugs is related to the necessity to maintain a certain concentration of drug in the system for adequate functioning of the therapy. After dosing, the concentration of a drug decreases over time. As the concentration falls below therapeutic levels, it is necessary to increase the concentration by administering the drug, usually a pill in this case, again. Two main problems arise from this situation. First, there is a varying concentration of the drug in the system, which can reduce drug effectiveness. Second, the regimen of re-administering the drug continuously is often not upheld, and patient compliance becomes an issue. In order to make medicine more convenient, thus increasing compliance and therefore the effectiveness of therapy, steps are being taken to increase the time and consistency with which ingested drugs deliver medicine.

There is active research in the area of slow- and controlled-release drugs administered orally, through either inhalation or ingestion. Time-release capsules have long been available on the commercial market, and some can provide 12 h of sustained drug concentrations before an additional dose is required. Most medications, however, require 3–6 doses per day to maintain therapeutic benefits. This is because the residency time in the small intestine, where most drug absorption occurs, is usually less than 4 h. Finding ways to increase the residency time of a drug in the digestive tract is key to oral-controlled release delivery.

18.1.3.1 Gastrointestinal tract. The transit time of food through the digestive tract is usually between 24 and 48 h. Most of this time is spent in the large intestine, an area with very limited capability to absorb drugs, and is therefore of little use for sustained drug delivery. Liquids can be removed from the digestive tract in less than 4 h. Medication in solid form, or taken with food passes through the upper gastro-intestinal (GI) tract (preceding the large intestine) in 5–8 h. The stomach has a small surface area, and with some exceptions, drugs are poorly absorbed in the stomach. Figure 18.8 shows that most absorption occurs in the small intestine, over a period of 3–5 h. In addition, the absorption of most drugs decreases as the drug travels through the small intestine, further decreasing the time of effective dosing. In order for controlled release of drugs in the GI tract to be practical, platforms must be developed for the release to occur from. The alternatives for platforms in the GI tract are limited to those that increase the time the drug is in the stomach releasing to the small intestine, or those that assist the drug to stay in the small intestine for an increased duration.

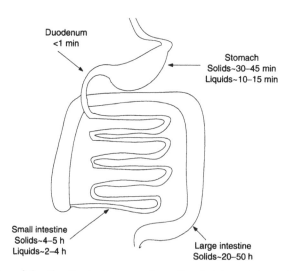

Figure 18.8. Outline of the digestive tract, as seen by orally administered drugs, showing the approximate residency times of liquids and solids in each of the regions. Almost all absorption of medicine occurs in the small intestine.

18.1.3.2 Intestinal platforms. Methods for creating platforms for the small intestine are limited. For a platform to reside in the small intestine, it must survive the environment of the stomach, and then selectively adhere to or be absorbed by the small intestine. Adhering systems have been attempted; however though there is promise, the specifics of the chemistry of the walls of the intestine need to be better understood for successful targeting. The employment of biodegradable microcapsules may provide an alternative solution. Desai *et al* (1996) performed a study in rats to determine the effects of size on particle uptake. It was found that particles 100 nm in size were absorbed through the submucosal layer of the intestine. Using this technique, it may be possible to encapsulate vaccines or drug matrices inside the polymers for release in the intestine.

18.1.3.3 Gastric retentive platforms. Increasing the residency time of a drug in the stomach may provide the necessary platform for controlled release. By making a delivery system that is difficult for the stomach to pass to the small intestine, a drug can be released over a period of time to the small intestine. Several methods have been proposed to serve this purpose. These are summarized here, but for a more detailed discussion see Hwang *et al* (1998).

Drug delivery systems proposed to increase residency time in the stomach employ such methods as floating or expanding the system in the stomach, sticking to the walls of the stomach, and even using magnets. With any of these platforms in place for drug delivery, a time-controlled release of medication can be accomplished and administration of the drug can be reduced in frequency, thus making the medication more convenient and effective.

Floating systems function by keeping the drug platform on top of the gastric fluids. Two basic methods can be employed. The platform can be made such that once in contact with the gastric fluids a decrease in density occurs, or the system can be administered in a low-density matrix. Systems that decrease in density do so in one of two ways. A gel-forming polymer with hydrophilic properties may be used, so that upon contact with the stomach fluids it expands and decreases in density. Alternatively, the drug matrix can be surrounded with gas-producing compounds and a semipermeable, expandable membrane. A simple method for this might be to use sodium bicarbonate as the gas-producing compound. When in contact with the stomach acids, the reaction produces carbon dioxide, filling the membrane with gas, and causing it to float.

Increasing the size of the drug delivery system can increase residency time. Large particles, greater than 15 mm in diameter, reside in the stomach for long periods. Optimally, the stomach reduces solids to 1–2 mm particles before passing them to the duodenum. Objects larger than 15 mm in diameter can reside in the stomach for several hours after primary digestion. Obviously, swallowing pills or capsules this big would be uncomfortable, so systems have been designed to expand or change shape after entering the stomach. One technique is to compress a geometric shape into a standard capsule. When in the stomach, the capsule would dissolve allowing the shape to expand to its previous shape. Another technique is to use hydrogels. These are compounds that polymerize into a volume several times (up to a thousand) their volume in the dry state. This system could be very effective, but currently they do not expand quickly enough to avoid being ejected from the stomach.

Attempts have been made to develop systems that stick to the walls of the stomach for an extended period. Several systems have been designed using poly(acrylic) acid. Although these systems have shown some improvement in residency time, so far they have not produced the results necessary. It is difficult to develop an adhesive that selectively sticks to the mucus layer. Adhesives without this property would be difficult to handle, unless encapsulated, and if encapsulated may react with the capsule.

Using magnets has been used to increase residency time in the stomach. Adding magnetic properties to the drug platform allows the position of the platform to be controlled by employing external magnets. Although this system is effective, the external magnet must be placed precisely and remain in position in order to keep the platform in place. This takes away the convenience of the long-release drug, making the system less practical.

Each of the systems above has its merit and downfall. All but the floating systems involve some extended contact with the stomach wall, which may cause irritation of the stomach lining, or even make the systems unusable by those with ulcers. The floating systems, however, have a limited ability to float, and are often ejected from the stomach even if they are floating properly. Oral sustained release remains attractive as a controlled-release method, and research will continue to improve the techniques to move toward effective systems.

18.1.4 Other noninvasive routes of administration

In addition to respiratory, transdermal and gastric delivery, research is being done in the areas of nasal delivery, sublingal delivery and rectal delivery. All of these routes have a commonality in that they involve absorption-enhancement techniques. Delivery through the nose and oral cavity show high delivery capability, however, it is limited mostly to small doses. In addition, with delivery directly through the nose and mouth, many patients find the taste of the medicine unpalatable. Rectal delivery, though effective in some cases, is also limited due to patient objections.

18.2 CONTROLLED-RELEASE DRUG DELIVERY

One of the biggest areas of interest in minimally invasive drug delivery is that of controlled and targeted delivery. Delivery to the specific region of the body needing treatment rather than delivering a drug to the whole body has distinct advantages. In the case where a specific region of the body needs to be treated with a drug, releasing the drug specifically in that area allows for a higher local concentration of medicine. This increases the therapeutic ratio of the drug in the necessary region, while decreasing the amount of drug needed overall. Systematic exposure to the drug is minimized. In many cases, a drug is ineffective below a certain threshold, but is toxic only at a slightly higher concentration. In such a case, localized delivery allows for therapies that might otherwise be deemed to put the patient at too great a risk.

18.2.1 Controlled-release delivery

Several methods and techniques have been developed to release drugs in a controlled manner, over a period of time, under certain conditions or in a specific region of the body. These range from slowly degrading microcapsules, to environmentally sensitive materials, to external stimulation of release. The basic components of a controlled-release system are a platform, a drug delivery mechanism, a rate controller, an energy source and a drug reservoir.

The platform is the mechanism that keeps the drug delivery system in a fixed position in the body. The platform can be passive, as is the case of an implant into muscle tissue. These implants remain in the muscle tissue and the muscle serves as the platform. Examples of active platforms are discussed in section 18.1.4. The devices designed to keep the drug system in the stomach or small intestine are the platform.

The drug delivery mechanism is the route of making the drug available to the body. This can be through passive diffusion, pumping or dissolution.

The rate controller is the mechanism by which the drug is released at a fixed rate. A semipermeable membrane can serve this purpose, as can a biodegradable drug matrix or an injection port.

Energy sources can be as simple as chemical reactions within the body. For example, in the case of a slowly dissolving drug matrix, the energy source is the fluids of the body responsible for dissolving the system.

The drug reservoir is where the drug is stored. This may take the form of an actual physical reservoir, or a drug matrix.

By varying the specifics of each of the components of the drug delivery system, many successful systems can be developed. They can be as simple as a drug matrix, which contains all the systems homogeneously, or as complex as an osmotic pump.

18.2.1.1 Preprogrammed delivery systems. Preprogrammed delivery systems are systems that are based on diffusion according to Fick's law of diffusion. Figure 18.9 shows that the basic system consists of a drug reservoir or matrix, surrounded by a rate-controlling membrane. As with any implanted system several criteria need to be met for safe delivery of drugs. The system must be biocompatable, so that it is not rejected or encapsulated by the body. The system must be nontoxic, and have nontoxic byproducts. The system must be able to be sterilized. The system must release drugs at a precise and consistent rate. The drug must not react with or degrade from contact with the controlling membrane. The system should degrade at a predictable rate, producing byproducts that are easily removed by the normal functions of the body.

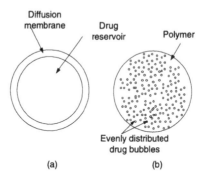

Figure 18.9. The two basic diffusion-controlled devices. (*a*) A device with a drug reservoir and membrane and (*b*) a polymer matrix.

Natural biodegradable polymers have been studied for drug delivery, however, they tend to be too expensive to be practical. Instead, synthetic biodegradable polymers have been developed to perform the function of rate-limiting barriers. Polymers such as poly(lactide), poly(glycolide) and poly(lactide-co-glycolide) have been used extensively in the development of implantable drug delivery systems and have been approved by the Food and Drug Administration for delivery of drugs (Jain *et al* 1998). Diffusion-controlled devices are designed such that they deliver the drugs through the membrane, after which, ideally, they degrade and are expelled by the body. Commercially available products, such as Norplant®, use diffusion from an implant to deliver long-term drug delivery. In the case of Norplant®, a contraceptive, sustained dose can occur for as long as five years. These implants must be removed surgically.

An alternative approach to encapsulation is the use of polymer matrices to deliver the drug. In these systems, a drug is uniformly dispersed throughout the delivery system, and enters the system by diffusion through the matrix and leeching from the matrix. Norplant-2® is a commercially available system employing this technique. The effectiveness of this system is comparable to the Norplant® system, however two implants, rather than six, are necessary for effective dosing over the five-year period.

18.2.1.2 Environmentally controlled delivery systems. Controlled-release systems that rely on interaction with the body to release the drug are classified as environmentally controlled systems. Examples of such systems include bioerodible systems and osmotic pumps. In their simplest form, these systems release drugs at a fixed rate to the system. It is possible to design the systems to react to changing body conditions, releasing the drug in varying concentrations depending on the systems' needs. Although such systems are sparingly researched in laboratories, with advances in material technologies, such devices such as glucose-sensitive release systems would greatly affect the treatment of chronic diabetes.

Bioerodible systems are drug matrices that do not deliver the drug through diffusion or leeching, but instead release the drug by slowly eroding over time. This system has the advantage that the implant by its functional nature need not be removed after treatment. A disadvantage of the system is that the erodible area decreases with time, making constant dosing a problem.

Osmotic pumps are nonerodible systems that encase a drug reservoir. Figure 18.10 shows that the system is a drug reservoir surrounded by an impermeable membrane. This impermeable membrane is surrounded by an osmotic agent, and is encased in a semipermeable membrane. When the system is implanted in the body, the osmotic agent draws in water through the semipermeable membrane. As the water is drawn its pressure is applied to the drug reservoir, which pushes some of the drug to the system.

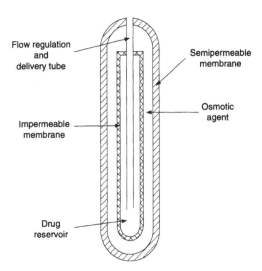

Figure 18.10. An osmotic pump is activated by the swelling of an osmotic agent that compresses a drug reservoir, pushing the drug out of the delivery system. Flow is regulated by the size and length of the exit tube.

One advantage of this type of system is the system can dose at a chosen rate, constant or variable, regardless of the environment of the implant or the nature of the drug being delivered. Currently available osmotic pumps can deliver a drug from 3 days to 4 weeks at rates ranging from 0.5 μL h^{-1} to 10 μL h^{-1}. Osmotic pumps are used primarily in the laboratory on animals.

18.2.2 Targeted-release delivery

Systematic dispersion of drugs can lead to side effects and reduce the therapeutic value of a treatment. In some cases, such as chemotherapy, it is advantageous to deliver drugs only to the area of the body affected in order to increase the localized dose while reducing systematic exposure. Alternatively, a drug can be designed so that systematic exposure does not result in systematic effects. These drugs, though presented to the whole system, selectively target a certain system or tissue and only affect directly the specific target.

18.2.2.1 Chemical and receptor targeting. Many tissues in the body are designed for specific molecules to attach to them, or are a receptor for a given molecular structure. By targeting these receptors, it is often possible to design a drug to act on a specific organ or tissue.

The body regulates many of its functions through ligand–receptor interactions. For example, blood sugar levels in the body are regulated by the insulin and glucagon receptors. The primary receptors for glucagon are on the liver. When blood glucose levels are low, the pancreas releases glucagon into the bloodstream. Upon receiving this signal, the liver cells release chemicals that break down glycogen (the stored form of glucose), and glucose is released into the bloodstream. When blood glucose levels are high, the pancreas releases insulin into the system. Insulin receptors exist on most cells in the body. When insulin is detected by the receptors, the storage of glucose is stimulated, removing glucose from the bloodstream. Dysfunction in the level of insulin in the body is called diabetes mellitus, or commonly diabetes.

In addition to providing communication within the body, ligand–receptor interactions can be used to convey molecules, ions or electrons to tissues in the body. For example, cholesterol is hydrophobic and is not soluble in water-based body fluids. In order for cholesterol to be distributed throughout the body, low-density lipoprotein (LDL) transports cholesterol throughout the body. Cells throughout the body have a LDL receptor. When LDL is recognized, the uptake of the molecule is triggered, delivering the cholesterol to the cell.

Receptors function by recognizing a ligand, attaching to it, and producing a physiological response. A ligand–receptor interaction is characterized by the following features: 'appropriate affinity, saturable number of sites, reversibility, stereospecificity, and temporal specificity' (Petty 1993). In addition to receptors, binding sites exist throughout the body where specific substances attach. These are called binding sites. Ligand–receptor interactions are complex, but can be understood at a basic level in the following manner. A receptor acts as a port for a ligand. The ligand contains a region that functions as a delivery address. When the receptor senses the appropriate address, it becomes a docking bay for the ligand. The receptor and ligand then act as a lock and key, and a ligand with the appropriate terminal sequences can attach to the receptor. When the ligand has attached to the receptor, it signals the cell to perform a function, such as the release of a chemical.

There are two main mechanisms that a drug can use in therapy involving receptors. The drug can artificially stimulate the response of the receptor as in insulin therapy, or it disrupts the actions of a receptor. The latter mode of action is common in the effects of many substances on the body. The effect of caffeine on sleep has been attributed to it blocking the actions of endogenous adenosine at its receptor.

Disrupting the action of a receptor can have a therapeutic benefit. The treatment of clinical depression has advanced rapidly due to the development of drug classes that affect the receptor response to dopamine and serotonin. Serotonin uptake inhibitors have been shown to have antidepressant effects, and are commonly used in the treatment of depression (Bonhomme and Esposito 1998). Long-term treatment with antidepressants induces adaptive changes in the cerebral monoaminergic systems. The basic action of this class of drugs is to desensitize the serotonin receptors, leading to an adaptive response in the brain. The adaptive response leads to a balancing of brain chemistry and often aids in reversing depression.

As mentioned with insulin, stimulating the response of a receptor can be used to treat a disease. In the case of insulin delivery, the drug is provided to, and used by, the entire system. In some instances, it is possible to target a specific organ or tissue by using receptors. This technique has been demonstrated as a possible treatment for cancer (Arap *et al* 1998, Huang *et al* 1997).

Modern treatment of cancer involves attacking the cancer directly, using chemotherapy, radiation therapy or surgical removal. Often the result of the treatment is incomplete destruction of the tumor, or in the case of chemotherapy, a system-wide attack. Chemotherapy often results in severe side effects, and destruction of healthy tissue. The goal of cancer therapy is to destroy the tumor without affecting healthy tissue. Recently, a new technique has been demonstrated in mice that may be able to accomplish this. Instead of attacking the tumor cells directly, techniques are being developed to cut off supply lines to the tumors.

Several studies, such as the two cited above, use receptor targeting to attack the blood vessels supplying tumors in mice. Huang *et al* (1997) use methods that block the process of blood vessel formation (angiogenesis) in the tumor and close off existing supply lines to the tumor by clotting them. The first approach involved targeting receptors involved in angiogenesis and blocking the ability of the tumor to generate new blood vessels. The second approach used the same targeting methods to deliver clotting agents to the vessels, sealing the supply of blood off from the tumor. Using this technique, they were able to stop tumor growth over the course of the study. Arap *et al* (1998) use the same targeting methods to deliver the anticancer agent dox-RGD-4C. With this treatment, more than 60% of the mice survived for six months, whereas the control mice all died after 70 days.

The advantages of targeting the vasculature of the tumor rather than the tumor itself include: the death of the cancer cells from starvation, the targeted receptors are in direct contact with the blood, and the endothelial cells are not mutated and would not likely mutate to acquire resistance to the therapy.

18.2.2.2 Externally controlled targeting. A novel approach to targeted drug release is to externally activate or stimulate the release of the drug. By providing a drug to the system and locally activating it, delivery of the drug remains simple, while localized action is preserved. The basis of this approach is to create delivery systems whose properties are affected by external stimuli. For example, it is possible to design drug matrices that release drugs at a slow rate under normal conditions, but in the presence of a magnetic field release at a greatly increased rate. Ultrasound and electric fields have also been used to stimulate release.

The magnetic and electric susceptible systems rely on embedding magnetic or ionic species, respectively, in a polymer matrix, along with the substance to be delivered. In the absence of an external field, these systems release at a slow rate or not at all. By applying oscillating fields, the structure of the matrix is modified and increased dosing occurs. This can be used to effect controlled pulsatile release of medicines, as well as moderately targeted delivery.

Ultrasound can also be used to release drugs, in either a controlled manner or a targeted manner. Price *et al* (1998) used ultrasound to deliver polymer microspheres and red blood cells to the muscle tissue of a rat. Delivery of agents to tissue often pose a difficulty in that the endothelial barrier of the blood cells is not always permeable to drug delivery vehicles. In order to deliver agents to the tissue, the group used ultrasound to locally rupture the microvessels in the tissue, allowing red blood cells and the microspheres to enter the tissue. Using this method they hypothesize that drug-bearing vehicles or engineered cells could be delivered to tissue.

18.2.2.3 Perfusions. Perfusion is the forced flow of a fluid through an organ or tissue. Traditionally in medicine, perfusions are used in open-heart surgery in the form of heart–lung machines. Although perfusions involve a surgical technique, in some cases they may be used to minimize damage to healthy tissue, thus classifying the technique as minimally invasive. In the area of drug delivery, perfusions are used to aid in chemotherapy.

As mentioned above, chemotherapy is often a destructive technique of therapy. Collateral damage to healthy tissue can be extensive. In addition, with healthy tissue at risk, the concentrations of chemotherapeutic agents must be kept below a certain level. By applying chemotherapy to the region containing the tumor, systematic toxicity can be decreased, while at the same time higher drug concentrations can be used to treat the tumor.

Perfusion is used to treat malignant tumors in the extremities and some internal organs. In this technique the area to be treated is isolated from the body, and chemotherapeutic agents are flowed through the area with the tumor. Figure 18.11 shows for treatment of melanoma in the leg that for example, the femoral artery and vein are cut and attached to an extracorporeal pump. A tourniquet is applied to the root of the leg to isolate it. The limb is then treated with chemotherapeutic agents through the perfusion machine. After the procedure, the artery is reattached and the incision closed (Papa *et al* 1996).

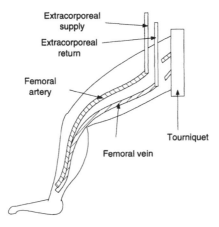

Figure 18.11. During a limb perfusion the limb is isolated from the body by a tourniquet, and chemotherapeutic agents are pumped through the limb from an extracorporeal pump. Treatment may last over an hour.

Less invasive procedures have been developed for treatment of organs such as the liver. For perfusion of the liver, Ku *et al* (1997) developed a single catheter technique. Figure 18.12 shows that using a catheter with two balloons spaced 5–6 cm apart, the blood supply to the liver is cut off. Through openings in the catheter between the balloons, the perfusion pump is operated, supplying treated blood to the liver. After up to 30 min of this technique, catheter balloons are deflated, and the catheter is removed. During the procedure, blood is filtered to remove the chemotherapeutic agents before blood is returned to the system.

These surgical techniques for chemotherapy result in effective treatment of the malignancies, while protecting the entire body from exposure to the chemotherapeutic agents. They have been demonstrated as relatively safe methods of treatment, and effective at controlling cancers.

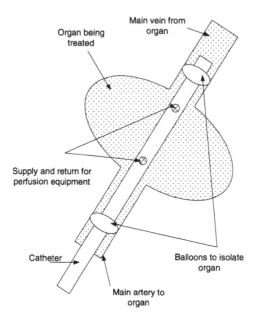

Figure 18.12. An organ perfusion for chemotherapy. Balloons are used to isolate the organ from the body so that the chemotherapeutic agents remain only in the organ being treated. The spacing of the balloons is modified to fit the size of the patient.

18.3 CONTROLLED-DOSE DELIVERY

One of the disadvantages of controlled-dose implants is the inability to adjust the dosing, and the difficulty in terminating treatment if necessary. In order to add an element of control to therapy, drug delivery systems with feedback or external control are needed.

18.3.1 Implantable systems and micropumps

For patients with chronic conditions that need continuous therapy, such as diabetics, an implantable delivery system simplifies drug delivery. In addition to implantable osmotic pumps, infusion pumps or implantable delivery chips may serve to simplify therapy.

Infusion pumps are devices implanted in the body to deliver drugs at a controlled rate. The systems in general have drug reservoirs that can be replenished. Implantable infusion pumps are used in the treatment of many chronic conditions, and the application of infusion pumps to assist in the therapy of dozens of conditions has been researched (Kohudic 1994). The primary research in this area is broadening the applications of the technology.

Recently, MEMS has provided a new pathway to implantable drug delivery. Santini *et al* (1999) used MEMS techniques to create a controlled-release microchip that contained 34 drug reservoirs, each capable of delivering 25 nl of solution. The reservoirs deliver the solution by electrochemical dissolution of thin gold membranes covering the reservoirs. This is accomplished by applying a 1 V signal to the gold film. In the presence of chlorine, the gold film dissolves, releasing the solution. They claim that, using their process, over 1000 reservoirs could be placed on a 17 mm × 17 mm chip.

Miniaturization and material science continue to improve the capability of implantable devices. Although most of the technology is still in the laboratory, the advances in techniques are creating more efficient and smaller pumps. Several of the newer systems are based on shape memory alloys, such as titanium–nickel. These alloys can deform, but return to an original shape when exposed to a certain temperature. Alloys such as these have been used on the macroscopic scale for some time, especially in orthodontics, where the alloys are used to help corrective devices maintain a certain shape. On smaller scales, shape memory alloys can be used as miniature actuators for micropumps (Benard *et al* 1998, Reynaerts *et al* 1997). By heating and cooling the metal, usually by Joule heating and thermal conduction for cooling, the alloys can be used to drive pumps. One method for doing this is to alternately heat two memory metal strips as complementary actuators.

18.3.2 Feedback systems

Ideally, a drug delivery system would be self-regulating. That is, a feedback system determines when a drug is needed in the system, and the delivery system is told to deliver it. Several of the implantable controlled-delivery systems attempt to do this by means of creating materials that have properties that are modified by the concentrations of certain molecules, the pH of the environment or other factors. Although no system exists in use that is entirely self-regulating, systems have been proposed and portions of them designed that could self-regulate. With the development of MEMS and integration of solid-state sensors, systems such as self-regulating insulin implants may be on the horizon.

PROBLEMS

18.1   List the advantages and disadvantages associated with respiratory delivery of medicine.

18.2   List the advantages and disadvantages associated with transdermal delivery of medicine.

18.3   Describe the major impediment to transdermal delivery and the methods that can be used to overcome this impediment.

18.4   Estimate the increase in iontophoretic flux resulting from a 20% drop in the resistance of the skin.

18.5   Compare ac versus dc iontophoresis.

18.6   Describe sonophoresis.

18.7   Compare the theoretical sonophoretic enhancement of hydrocortisone (P_p = 1.3×10^{-4} cm h^{-1}, $K_{o/w}$ = 40) and lidocaine ($P_p = 4 \times 10^{-3}$ cm h^{-1}, $K_{o/w} = 200$).

18.8   List the advantages of controlled-release drug delivery.

18.9   List the advantages of targeted drug delivery.

REFERENCES

Arap W, Pasqualini R and Ruoslahti E 1998 Cancer treatment by targeted drug delivery to tumor vasculature in a mouse model *Science* **279** (5349) 377–80

Banga A K and Prausnitz M R 1998 Assessing the potential of skin electroporation for the delivery of protein- and gene-based drugs *Trends Biol.* **16** 408–12

Benard W L, Kahn H, Heuer A H and Huff M A 1998 Thin-film shape-memory alloy actuated micropumps *J. Microelectromech. Syst.* **7** 245–51

Bonhomme N and Esposito E 1998 Involvement of serotonin and dopamine in the mechanism of action of novel antidepressant drugs: a review *J. Clin. Psychopharmacol.* **18** 447–54

De Boer A G 1994 *Drug Absorption Enhancement: Concepts, Possibilities, Limitations and Trends* (Langhorne, PA: Harwood)

Desai M P, Labhasetwar V, Amidon G L and Levy R J 1996 Gastrointestinal uptake of biodegradable microparticles: effect of particle size *Pharmaceut. Res.* **13** 1838–45

Duck F A, Baker A C and Starritt H C 1998 *Ultrasound in Medicine* (Bristol: Institute of Physics)

Florence A T and Salole E G 1990 *Routes of Drug Administration* (Boston, MA: Wright)

Gallo S A, Oseroff A R, Johnson P G and Hui S W 1997 Characterization of electric-pulse-induced permeabilization of porcine skin using surface electrodes *Biophys. J.* **72** 2805–11

Hadgraft J and Pugh W J 1998 The selection and design of topical and transdermal agents: a review *J. Invest. Dermatol.* **3** 131–5

Hari P R, Chandy T and Sharma C P 1996 Chitosan/calcium alginate microcapsules for intestinal delivery of nitrofuratoin *J. Microencapsulation* **13** 319–29

Henry S, McAllister D V, Allen M G and Prausnitz M R 1998 Micromachined needles for the transdermal delivery of drugs *Proc. MEMS 98 IEEE 11th Annual Int. Workshop Micro Electro Mech. Syst.* pp 494–8

Huang X, Molema G, King S, Watkins L, Edgington T S and Thorpe P E 1997 Tumor infarction in mice by antibody-directed targeting of tissue factor to tumor vasculature *Science* **275** (5299) 482–4

Hwang S J, Park H and Park K 1998 Gastric retentive drug-delivery systems *Crit. Rev. Therapeut. Drug Carrier Syst.* **15** 243–84

Inoue N, Kobayashi D, Kimura M, Toyama M, Sugawara I, Itoyama S, Ogihara M, Sugibayashi K and Morimoto Y 1996 Fundamental investigation of a novel drug delivery system, a transdermal delivery system with jet injection *Int. J. Pharmaceut.* **137** (21) 75–84

Jain R, Shah N H, Malick A W and Rhodes C T 1998 Controlled drug delivery by biodegradable poly(ester) devices: different preparative approaches *Drug Dev. Indust. Pharm.* **24** 703–27

Kefalides P T 1998 New methods for drug delivery *Ann. Intern. Med.* **128** 1053–5

Kohudic M A 1994 *Advances in Controlled Delivery of Drugs* (Lancaster: Technomic)

Ku Y, Fukumoto T, Masahiro T, Iwasaki T, Maeda I, Kusunoki N, Obara H, Sako M, Suzuki Y, Kuroda Y and Saitoh Y 1997 Single catheter technique of hepatic venous isolation and extracorporeal charcoal hemoperfusion for malignant liver tumors *Am. J. Surg.* **173** 103–9

Langkjaer L, Brange J, Grodsky G M and Guy R H 1998 Iontophoresis of monomeric insulin analogues *in vitro*: effects of insulin charge and skin pretreatment *J. Controlled Release* **51** 47–56

Li S K, Ghanem A H, Peck K D and Higuchi W I 1997 Iontophoretic transport across a synthetic membrane and human epidermal membrane: a study of the effects of permeant charge *J. Pharm. Sci.* **86** 680–9

Li S K, Ghanem A H, Peck K D and Higuchi W I 1998 Characterization of the transport pathways induced during low to moderate voltage iontophoresis in human epidermal membrane *J. Pharm. Sci.* **87** 40–48

Li S K, Ghanem A H, Peck K D and Higuchi W I 1999 Pore induction in human epidermal membrane during low to moderate voltage iontophoresis: a study using ac iontophoresis *J. Pharm. Sci.* **88** 419–27

Mitragori S, Blankschtein D and Langer R 1996 Transdermal drug delivery using low-frequency sonophoresis *Pharmaceut. Res.* **13** 411–20

Mitragori S, Blankschtein D and Langer R 1997 An explanation for the variation of sonophoretic transdermal transport enhancement from drug to drug *J. Pharmaceut. Sci.* **86** 1190–91

Papa M Z, Klein E, Karni T, Koller M, Davidson B, Azizi E and Ben-Ari G 1996 Regional hyperthermic perfusion with cisplatin following surgery for malignant melanoma of the extremities *Am. J. Surg.* **171** 416–20

Petty H R 1993 *Molecular Biology of Membranes* (New York: Plenum)

Price R J, Skyba D M, Kaul S and Skalak T C 1998 Delivery of colloidal particles and red blood cells to tissue through microvessel ruptures created by targeted microbubble destruction with ultrasound *Circulation* **98** 1264–7

Reynaerts D, Peirs J and Van Brussel H 1997 An implantable drug-delivery system based on shape memory alloy micro-attenuation *Sensors Actuators A* **61** 455–62

Santini J T, Cima M J and Langer R 1999 A controlled-release microchip *Nature* **397** (6717) 335–38

Venkatraman S and Gale R 1998 Skin adhesives and skin adhesion. 1: Transdermal drug delivery systems *Biomaterials* **19** 1119–36

Wu J, Chappelow J, Yang J and Weimann L 1998 Defects generated in human stratum corneum specimens by ultrasound *Ultrasound Med. Biol.* **24** 705–10

Zakzewski C A, Wasilewski J, Cawley P and Ford W 1998 Transdermal delivery of regular insulin to chronic diabetic rats: effect of skin preparation and electrical enhancement *J. Controlled Release* **50** 267–72

INDEX